Transecology

There is a growing recognition of the importance of transgender perspectives about the environment. Unlike more established approaches in the environmental humanities and queer studies, transecology is a nascent inquiry whose significance and scope are only just being articulated. Drawing upon the fields of gender studies and ecological studies, contributors to this volume engage major concepts widely used in both fields as they explore the role of identity, exclusion, connection, intimacy, and emplacement to understand our relationship to nature and environment.

The theorists and ideas examined across multiple chapters include Stacy Alaimo's notion of "trans-corporeality" as a "contact zone" between humans and the environment, Timothy Morton's concept of "mesh" to explore the interconnectedness of all beings, Susan Stryker's notion of trans identity as "ontologically inescapable," Catriona Mortimer-Sandilands and Bruce Erickson's history of the development of queer rural spaces, Judith Butler's analysis of gender as "performative"—with those who are not "properly gendered" being seen as "abjects"—and Julia Serano's contrasting rejection of gender as performance.

Transecology: Transgender Perspectives on Environment and Nature will be of great interest to scholars, graduate students, and advanced undergraduates in transgender studies, gender studies, ecocriticism, and environmental humanities.

Douglas A. Vakoch is President of METI, dedicated to Messaging Extraterrestrial Intelligence and sustaining civilization on multigenerational timescales. As Director of Green Psychotherapy, PC, he helps alleviate environmental distress through ecotherapy. Six of his earlier books explore ecofeminism and ecopsychology.

Routledge Studies in Gender and Environments

Series Editor: Professor Susan Buckingham, an independent researcher, consultant and writer on gender and environment related issues.

With the European Union, United Nations, UN Framework Convention on Climate Change, and national governments and businesses at least ostensibly paying more attention to gender, including as it relates to environments, there is more need than ever for existing and future scholars, policy makers, and environmental professionals to understand and be able to apply these concepts to work towards greater gender equality in and for a sustainable world.

Comprising edited collections, monographs and textbooks, this new *Routledge Studies in Gender and Environments* series will incorporate sophisticated critiques and theorizations, including engaging with the full range of masculinities and femininities, intersectionality, and LBGTIQ perspectives. The concept of 'environment' will also be drawn broadly to recognize how built, social and natural environments intersect with and influence each other. Contributions will also be sought from global regions and contexts which are not yet well represented in gender and environments literature, in particular Russia, the Middle East, and China, as well as other East Asian countries such as Japan and Korea.

Titles in this series include:

Ecological Masculinities
Theoretical Foundations and Practical Guidance
Martin Hultman and Paul M. Pulé

Gendered Approaches to Spatial Development in Europe
Perspectives, Similarities, Differences
Edited by Barbara Zibell, Doris Damyanovic and Ulrike Sturm

Gender and Religion in the City
Women, Urban Planning and Spirituality
Edited by Clara Greed

Transecology
Transgender Perspectives on Environment and Nature
Edited by Douglas A. Vakoch

Transecology

Transgender Perspectives on Environment and Nature

Edited by Douglas A. Vakoch

Routledge
Taylor & Francis Group
LONDON AND NEW YORK

earthscan
from Routledge

First published 2020
by Routledge
2 Park Square, Milton Park, Abingdon, Oxon OX14 4RN

and by Routledge
52 Vanderbilt Avenue, New York, NY 10017

Routledge is an imprint of the Taylor & Francis Group, an informa business

British Library Cataloguing-in-Publication Data
A catalogue record for this book is available from the British Library

Library of Congress Cataloging-in-Publication Data
A catalog record has been requested for this book

ISBN: 978-0-367-08651-0 (hbk)
ISBN: 978-0-429-02381-1 (ebk)

Typeset in Times New Roman
by Wearset Ltd, Boldon, Tyne and Wear

To Joe Castrovinci

Contents

Figures

Contributors

Nicole Anae graduated from Charles Sturt University with a BEd and DipT before earning her PhD through the Faculty of English, Journalism and European Languages at the University of Tasmania. *Her research interests* typically explore historical and contemporary encounters between the self, embodiment, and culture with a view to examining issues of identity and performance. She is Senior Lecturer in Literary and Cultural Studies at Central Queensland University and has also worked at the University of South Australia (Mawson Lakes), the University of the South Pacific (Laucala Bay Campus in Suva, Fiji), Charles Sturt University (Wagga Wagga), and the University of Southern Queensland (Springfield).

Anna Bedford teaches Women, Gender, and Sexuality Studies at St. Mary's College of Maryland. She earned her PhD in Comparative Literature from the University of Maryland, College Park, and her MA in English Literature from the University of Edinburgh, Scotland. Bedford has published several articles on environmentalism, feminism, gender, and science fiction. She is currently co-editing an anthology of early women science fiction writers from the pulp and magazine era of the 1920s to the 1950s.

Finn Enke is a Professor of History, Gender and Women's Studies, and LGBTQ+ Studies at University of Wisconsin, Madison. Enke is author of *Finding the Movement: Sexuality, Contested Space, and Feminist Activism* (2007) and editor of *Transfeminist Perspectives in and Beyond Transgender and Gender Studies* (2012). They are currently working on a graphic memoir entitled *With Finn and Wing, Growing Up Amphibious in a Nuclear Age.*

Mat Fournier is an Assistant Professor in the Department of Modern Languages and Literature at Ithaca College. He obtained his PhD in Comparative Literature with a minor in Gender Studies at the Université Paris 8—Vincennes-Saint-Denis, France, in 2015. His main area of research is modern and contemporary European literature and his methods of analysis are informed by transgender studies, queer theory, and continental philosophy, particularly the common work of Gilles Deleuze and Félix Guattari.

He is particularly interested in the articulation of gender and political theory. His dissertation, *Queer Geophilosophies: Homosexuality, Transsexuality and Exile in Europe in the Interwar Years*, focused on the relationships between gender disruption, political subjectivities, and first-person writing in the heavily polarized geopolitical context of the 1930s. He is currently working on a book tentatively entitled *Sexing the Political: Gender Crisis and the European Civil War*. His articles have recently appeared in journals such as *Transgender Studies Quarterly*, *Journal of Postcolonial Writing*, and *L'Esprit Créateur*.

Gail Grossman Freyne, LLB (Melbourne), PhD (Dublin), MFTAI (Ireland) is a family therapist, mediator, and author who has been working in private practice since 1987 as a founding member of the Family Therapy and Counselling Centre, Dublin, Ireland. Prior to this, Freyne, who grew up in Melbourne, Australia, practiced for many years as a solicitor in that country and in New Orleans. Her published books and articles have considered gender and power in marriage and women's situation in the Catholic Church. At present she is working on a new book that considers the feminist themes in the detective stories of Dorothy L. Sayers. Freyne has taught both the Feminism and Family Therapy and the Homosexuality in Therapy modules for the Clanwilliam Institute, Dublin. For many years, she taught the Ecofeminism module for the MA in Theology and Ecology for Lampeter University, Wales. She has also lectured at the Von Hugel Institute, St. Edmunds College, Cambridge, England, and is a regular guest on radio and TV programs dealing with feminist issues. She also serves on the Board of the Joan D. Chittister Foundation for Contemporary Spirituality in the United States. Freyne is the author of "Better to Love than to Fade Away" in *Spiritual Questions for the Twenty First Century* (Orbis, 2001) and *Care, Justice and Gender: A New Harmony for Family Values* (Veritas, 2006).

Greta Gaard is Professor of English and Coordinator of the Sustainability Faculty Fellows at University of Wisconsin—River Falls. Her work emerges from the intersections of feminism, environmental justice, queer studies, and critical animal studies, exploring a wide range of issues ranging from interspecies justice, material perspectives on fireworks and space exploration, postcolonial ecofeminism, and the eco-politics of climate change. Author or editor of 6 books and over 80 refereed articles, Gaard's most recent volume is *Critical Ecofeminism*. She serves as a manuscript reviewer for university presses with environmental lists, a reviewer for feminist journals such as *Hypatia, Feminist Media Studies, Feminism & Psychology*, and for environmental, sustainability-focused, and animal studies journals. Her creative nonfiction eco-memoir, *The Nature of Home* (Arizona University Press, 2007), has been translated into Chinese and Portuguese.

Katie Hogan is a Professor of English and Faculty Affiliate of women's and gender studies at the University of North Carolina, Charlotte. Her book

Women Take Care: Gender, Race, and the Culture of AIDS was a finalist for the CGS Gustave O. Arlt Award in the Humanities. Her current project, *Imagining Climate Futures Through Queer and Trans Literature*, explores climate future narratives rooted in gender, sexual, and economic justice.

Peter I-min Huang (黃逸民) is a Professor Emeritus in the English Department of Tamkang University, Taiwan. His areas of teaching and research focus on ecofeminism, Chinese sci-fi, cli-fi, GLBTQ, postcolonial ecocriticism, and posthumanism studies. During his service as Chair of the English Department (2007–2011), he was the conference-organizing Chairperson for *The Fourth Tamkang International Conference on Ecological Discourse* (May 23–24, 2008) and *The Fifth Tamkang International Conference on Ecological Discourse* (December 17–18, 2010). He is the author of *Linda Hogan and Contemporary Taiwanese Writers: An Ecocritical Study of Indigeneities and Environment* (Lexington, 2015). Huang's other publications include journal articles (in *Neohelicon*; *Journal of Poyang Lake*; and *CLCWeb: Comparative Literature and Culture* journal); and book chapters (in *Literature and Ecofeminism: Intersectional and International Voices; Ecocriticism in Taiwan: Identity, Environment, and the Arts*; *East Asian Ecocriticisms: A Critical Reader*; and *The Future of Ecocriticism: New Horizons*).

Hee-Jung Serenity Joo is an Associate Professor in the Department of English, Film, and Theatre at the University of Manitoba in Winnipeg, Canada. Her research and teaching interests are in ethnic American literature, critical race studies, and speculative fiction. She has published essays on comparative racial passing, the racial politics of disaster fiction and film, and the limits of human rights discourse in journals that include *MELUS*, *American Studies*, *Journal of Transnational American Studies*, and *Literature Interpretation Theory*. Her current projects include a manuscript on race and/in the Anthropocene, and a collaborative endeavor on representations of the rural in Asian North American literature.

Julia Kuznetski (Tofantšuk) is an Associate Professor of British Literature at the School of Humanities of Tallinn University, Estonia. Her MA thesis was on gender issues in Angela Carter's writing and her PhD dissertation (2007) was on the construction of identity in the fiction of contemporary British women writers, focusing on Eva Figes, Jeanette Winterson, and Meera Syal. She has published articles and delivered papers on gender, identity, trauma, Herstory, the Holocaust, ecocritical concerns in British literature, post-colonial and migrant British literature, and comparative Estonian and British literature (for example, through Palgrave Macmillan, Peter Lang WV, VW Trier, and *Interlitteraria*). Her recent research and teaching interests include literature by women, gender studies, ecocriticism, ecofeminism, diasporic identities, art history, women and landscape in art, and comparative literature. She has developed and taught MA courses focusing on ecocritical considerations of twentieth-century literary texts, especially novels by Graham

Swift, Anita Desai, and Monique Roffey, as well as a theoretical PhD course on ecocriticism, place theory, and ecofeminism, and co-supervised PhD theses that employ a postcolonial, ecofeminist, and material ecocritical perspective. Kuznetski is a member of the Association for the Study of Literature and Environment (ASLE), the European Society for the Study of English (ESSE), the Sylvia Townsend Warner Society, and Estonian Women's Studies Center.

Elizabeth Parker attained her first-class degree in English Literature from Royal Holloway, University of London, before going on to do an MPhil in Popular Literature at Trinity College Dublin, in which she was awarded a distinction. As an Irish Research Council (IRC) scholar at Trinity, she was awarded her PhD. She has lectured in English Literature and Popular Culture at a number of universities across the UK and Ireland. She is the author of *The Forest and the EcoGothic: The Deep Dark Woods in the Popular Imagination* and is the founding editor of the journal *Gothic Nature: New Directions in Ecohorror and the EcoGothic*. Her research interests include the intersections between popular culture, horror, the Gothic, and the environmental humanities.

Nicole Seymour is an Associate Professor in the Department of English, Comparative Literature, and Linguistics at California State University, Fullerton (CSUF). She works in the environmental humanities, asking how literature and other cultural forms—from documentary film to stand-up comedy—mediate our relationship to environmental crisis. Seymour earned her PhD in English Literature at Vanderbilt University, and before coming to CSUF she held positions at the University of Louisville, the University of Arkansas, and the Rachel Carson Center for Environment and Society in Munich, Germany. Her first book, *Strange Natures: Futurity, Empathy, and the Queer Ecological Imagination*, won the 2015 scholarly book award from the Association for the Study of Literature and Environment (ASLE). Her forthcoming book, *Bad Environmentalism: Affect and Dissent in the Ecological Age* (University of Minnesota Press), recuperates irreverent and anti-sentimental expressions of environmentalism. Seymour enjoys working with students on research and public activism projects such as Climate Change Theatre Action. She currently serves as Vice President of ASLE and as an Area Team Coordinator for the UC-CSU Knowledge Action Network for Transformative Climate and Sustainability Education and Action.

Wibke Straube is a Senior Lecturer at the Centre for Gender Studies at Karlstad University (KAU), Sweden. They earned their PhD in Gender Studies from Linköping University, Sweden. Straube's work has for many years been situated in the field of Transgender Studies. Their most recent research is situated in feminist environmental humanities and interrogates questions of intimacy, transing bodies, and transecological aesthetics in visual media. Generally, their research and teaching interests focus on visual cultures, affect theory, intersectional feminist theory, ecology, and feminist posthumanities. Straube has a long-standing attachment to queerfeminist and transfeminist activism reflected also in their background in transdisciplinary Gender Studies. Their PhD dissertation, *Trans Cinema and its Exit Scapes. A Transfeminist Reading of Utopian Sensibility and Gender Dissidence in Contemporary Film* (Linköping University Press, 2014), explored utopian, cinematic storytelling and the worlding potential of song, dance, and dream scapes in trans cinema. They currently hold a teaching as well as research coordinator position for the Swedish Gender Studies network GEXcel International Collegium where they organize academic events, among others the trans studies research residencies for the Centre for Gender Studies at KAU. They are affiliated with the *Posthumanities Hub* based at the Royal Institute of Technology (KTH).

Susan Stryker earned her PhD in United States History from UC Berkeley and later held a postdoctoral fellowship in Sexuality Studies at Stanford University. She has been a distinguished visiting faculty member at Harvard University, University of California Santa Cruz, Simon Fraser University in Vancouver, and Macquarie University in Sydney. Before joining the Faculty of the University of Arizona in 2011 to direct the Institute for LGBT Studies, Stryker was Associate Professor of Gender Studies at Indiana University. She is a multiple award-winning author, editor, and filmmaker whose credits include the Emmy-winning documentary *Screaming Queens: The Riot at Compton's Cafeteria*, the massive, two-volume *Transgender Studies Reader*, the introductory textbook *Transgender History*, and popular non-fiction works such as *Queer Pulp: Perverse Passions from the Golden Age of the Paperback* and *Gay by the Bay: A History of Queer Culture in the San Francisco Bay Area.* Stryker also worked for several years as Executive Director of the GLBT Historical Society in San Francisco.

Katherine Thorsteinson is an Assistant Professor of Media Studies at St. Thomas University's English Department in Fredericton, New Brunswick, Canada. She received her PhD in English at Cornell University, where she wrote her dissertation *Narratives of Disposability: Race and Ecology in Contemporary Media.* She completed her BA (Hons) in English and philosophy at the University of Toronto, followed by her MA in English at the University of Manitoba. She has written on African American folklore, the Black Diaspora in Canada, queer theory, and collaborative scholarship in the humanities. Her articles have appeared in journals that include *Criticism*, *Diaspora*, *Modern Fiction Studies*, *Poetics Today*, *Transformations*, and *Writing from Below*. Her research interests pertain largely to issues of captivity and escape, from African American oral culture and slave narratives, to the chronopolitics of contemporary prison memoirs, to theories of gender and sexuality.

Douglas A. Vakoch is President of METI, a non-profit research and charitable organization dedicated to Messaging Extraterrestrial Intelligence and helping sustain civilization on multigenerational timescales. His edited and co-edited books include *Ecofeminism and Rhetoric* (2011), *Feminist Ecocriticism* (2012), *Ecopsychology, Phenomenology, and the Environment* (2014), *Women and Nature?* (2017), *Ecofeminism in Dialogue* (2017), and *Literature and Ecofeminism* (2018). Vakoch is general editor of two environmental book series: Ecocritical Theory and Practice, with over 60 books published to date, and the recently launched Environment and Society. He is Director of Green Psychotherapy, PC, a private practice in the San Francisco Bay Area that helps people overcome psychological distress through ecotherapy and sustainable practices. He is also a Professor Emeritus of Clinical Psychology at the California Institute of Integral Studies (CIIS).

Foreword

Susan Stryker

A number of years ago I had the opportunity to hike for a few days in some of the geologically newest land on the face of the planet: the almost uninhabitable volcanic desert of Iceland's interior highlands. Almost uninhabitable, because there I was.

The land arose from the water only 16 million years ago, accreting upward from the seabed fissure where the Eurasian and North American tectonic plates pull apart, and its surface had been scraped utterly clean during the last ice age just over 10,000 years ago. Even now it is free of snow cover for only a few months a year, and the ground is so porous that meltwater soaks rapidly through it. Except in a few scattered oases around hot springs and glacial rivers, there is no vegetation other than the lichens, fungi, and moss that cling tight and low to the rocks. Therefore there is no soil to speak of, and therefore not much of a food chain—no vascular plants, no insects, no terrestrial herbivores or four-legged predators. A raven or sea eagle might fly high overhead, without much incentive to land. The Icelandic highlands are for the most part a place of light and darkness, of atmosphere, of ice and stone rather than biotic life.

If the old stories are to be believed, the highlands are home to giants, elves, trolls, and fairies. (And why must the nonhuman be conceived of as natural rather than supernatural, subhuman rather than superhuman, particularly if we are to honor the wisdom of indigenous cosmologies?) And us. We are there in the Icelandic highlands, we humans and the companionate horses and dogs that historically have traveled with us in the little less than 1,200 years that our kind has lived on that land, periodically traversing a terrain too barren for settlement, yet perfectly able to be passed through as we carry our food and shelter and survival skills with us. Because I am there, and because I am trans, this is a transecology. It is an environment capable of accommodating naturally techno-cultural me and my kind and our tools and fuel, even as it accommodates so little else of life.

"Nature:" the poet Oliver Bendorf has written, in his own seductive meditation on tranecologies, "if you lived here you'd be home by now" (2014, 136–7). But we do live here, we transfolk who are home/not home in nature.

Trekking for hours in utter solitude and silence across the black lava sands of the Icelandic interior affords a powerful perspective on what has come to be called the Anthropocene. On the one hand, the term offers a name for the astonishingly extremophilic creatures we are, whose numbers have proliferated with such virulent ferocity across the globe, into benthic and extraterrestrial environments so harsh that they sustain only us and the microbes, with such consequence for other forms of life that we might as well name a geological era for the species whose mode of life is wreaking change on a geological scale. And on the other hand, do we really need a term that once again centers an anthropic exceptionalism at the precise moment in planetary history when Anthropos clings by its frail fingernails on the precipice of extinction, and threatens to take a good chunk of the kingdom of life with it as it falls (Moore 2016)?

Perhaps all that really matters is whether words and concepts like "Anthropocene" and "life" help us parse the distinctions between the natural history of the technological modes of our species-life, the hierarchizing of life according to the historically specific forms of violence called colonialism and racism that produce the fictive figure known by the name of Man, and our ability to create that unevenly distributed and fluidly boundaried capacitation of life we call the human. Our life is always a trans-life, that simultaneously occupies the ranked and contested intersections of Man, the human, the species of homo sapiens, and the animate nonhuman. It is a perpetually precarious life, a life always at risk of death and subjugation, a fugitive life that needs an elsewhere to sustain itself. Our concepts of life, and of life's spread through time and space, should help move us toward a better ethics—a better life.

I write these thoughts on transecology while sitting on a sandy beach, back against a crumbling cliff, near the mouth of the Onkaparinga River in South Australia where my friend Nikki Sullivan now lives. Spaces become places through our interactions and affective connections with them, and Australia will always be Australia for me through the work I have done there over the past decade with the Somatechnics Research Network, which emerged from a series of conferences organized by Sullivan beginning in 2003. The network is a transnational group of cultural studies scholars convened in recognition of the insights that embodiment is always technologized embodiment, that soma and techné never really have an "and" between them, that technologies are never prosthetically added to a naturally pre-existing body shaped in the absence of technology, that bodies and milieus always co-become with and through each other by particular means and according to particular manners: we are all transcorporeal technobodies (Sullivan and Murray 2009). Saying otherwise only favors those who can perform the sleight of hand that makes the technology of their becoming disappear into the ambient background. The trick is to remember that not all technologies are of human origin; we are acted-upon, and are instruments of other wills, as much if not more than we act as tool-wielding agents.

Somatechnics has one strong root in the white Australia feminist tradition of corporeal philosophy, including the work of Rosalyn Diprose. In *The Bodies of Women: Ethics, Embodiment, and Sexual Difference*, Diprose writes that ethics is not a set of abstract moral principles, but rather is always embodied in relation to place. The word *ethics* is derived from the Greek ethos, meaning character, habitat, and dwelling (and thus a close kin to oikos, a dwelling place or household, that provides the root for the prefix eco- in ecology). Dwelling, Diprose reminds us, is both a noun and a verb: it is a place to which one returns, as well as the practice of abiding there—both a habitat and a habitual occupation of a given place. One's habitual way of life constitutes one's character or ethos, which is not given in advance but rather produced through bodily acts, whose qualities are governed by the habitat in which they transpire. "From this understanding of ethos," Diprose concludes, "ethics can be defined as the study and practice of what defines one's habitat, or as the problematic of the constitution of one's embodied place in the world" (1994, 18–19). Ethics, in other words, is inherently transecological, and must be attuned to histories—and emergences—of embodied difference in all its forms, including those (perhaps especially those) that problematize the nature/culture duality.

I have been asked to write this foreword, I suppose, because more than a quarter-century ago, in an essay that expressed my affinity and kinship with Frankenstein's creature on the basis of my "unnatural" transsexual embodiment, I critiqued notions of Nature that were but bedeviling lies, fictions that cloaked, beneath the pretense of inevitability, the machinations of social power so detrimental to my life and to the lives of those like me. The enemy of my nature is a Nature that is home to Man, but not to me. I asserted then my sense of life as being filled with monstrous potential in which I acknowledged my "egalitarian relationship with nonhuman material being" (Stryker 1994, 240). It hurts, and is dangerous, to be dehierarchized, to lose human status by falling outside of norms and thereby being subjected to violence, but decentering the tangled webs of trans-huManinimality nevertheless offers a better ethical starting place for enacting our relationship to Being than trying to prop up a spurious anthropocentric privilege. This is where transecology begins for me.

Most of what is, is nothing, the feminist physicist Karen Barad (2015, 394–6) reminds us: galaxies, like atoms, are teeming voids filled with quantum virtualities that play out the possibilities of their potential becomings in the darkness of unbeing. Within such an ontology, trans*life and transecologies are not reducible to physicalities, however much we take into account the queer generativity of their originary technicity. Life is more than the biotic, which splits and recombines the elements necessary for the material reproduction and passing on of its forms and processes and habitual ways of being. Life is an even more fundamental process of grasping virtual possibilities immanent in the organization of the world but as yet unrealized within it, and transness is the techné for bringing forth those virtualities across scales and registers of Being. Transecologies are vast—vaster even than the dimensions explored in this important anthology—and encompass the conditions of life both actual and potential, conditions that are both needed and desired, and which may yet come to be.

References

Barad, Karen. 2015. "TransMaterialities: Trans*/Matter/Realities and Queer Political Imaginings." *GLQ* 21(2–3): 387–422.

Bendorf, Oliver. 2014. "Nature." *TSQ* 1(1–2): 136–7.

Diprose, Rosalyn. 1994. *The Bodies of Women: Ethics, Embodiment, and Sexual Difference*. London: Routledge.

Moore, Jason W., ed. 2016. *Anthropocene or Capitalocene: Nature, History, and the Crisis of Capitalism*. Oakland, CA: PM Press.

Stryker, Susan. 1994. "My Words to Victor Frankenstein Above the Village of Chamounix: Performing Transgender Rage." *GLQ* 1(4): 237–54.

Sullivan, Nikki, and Samantha Murray, eds. 2009. *Somatechnics: Queering the Technologisation of Bodies*. London: Routledge.

Preface

Greta Gaard

A field is founded when questions are asked that cannot be answered within the purview of existing disciplines. "What is it?" is the question asked about Jess Goldberg, hero of Leslie Feinberg's *Stone Butch Blues*, the book that in 1993 brought trans* identities to the center of feminist, queer, and literary conversations. That pivotal question—What is it?—signals the fact that Jess' identity exceeds the knowledge boundaries of gender dualisms, introducing a challenge to those restrictions and proposing a new perspective on identity, sexuality, and politics. In the time period of the novel, the 1950s and 1960s, that challenge a new perspective were not only unwelcome but were seen as punishable through physical, sexual, and economic violence.[1] Although Jess (and Feinberg) explored gender reassignment surgery and chose to identify as simply "butch," remaining in the liminal space that might now be called genderqueer, the novel profiles several transgendered characters, along with a trans* history and community that was later documented in Feinberg's *Transgender Warriors* (1997).

Transecology's genealogy is simultaneously ecocritical, queer, feminist, critically reconstructive and trans*disciplinary. As Eva Hayward and Jami Weinstein (2015, 196, 197) explain,

> trans* is not a thing or being, it is rather the processes through which thingness and beingness are constituted. In its prefixial state, trans* is prepositionally oriented—marking the with, through, of, in, and across that make life possible … trans* troubles ontologized states.

On this view, ecocriticism has from the beginning invoked a trans*disciplinary methodology in reading across literary, narrative, and cultural genres to create the grounds among, with, and through its multiple standpoints, and is thus well positioned for developing more openly transgendered critiques.

Ecocriticism's origins began a year after the publication of *Stone Butch Blues*, when literary critics meeting at the Western Literature Association decided to launch a new organization, based on what they saw as a lacuna in literary studies. From Annette Kolodny's feminist literary criticism in *The Lay of the Land* (1975), they knew that European explorers had attributed gender to the "new world," feminizing environments in ways that authorized environmental "rape" and appropriation of "virgin forests." With Kolodny, these new environmental literary critics also asked questions that exceeded the boundaries of literary criticism, beginning with the question, "Why is writing about nature and environment excluded from the high canon of literary traditions?" Although their initial focus was on US nature writing by Euro-American authors such as Sarah Orne Jewett (*Country of Pointed Firs*), John Muir (*My Life in the Sierras*), Henry David Thoreau (*Walden*), Aldo Leopold (*A Sand County Almanac*), Rachel Carson (*Silent Spring*), and first articulated in *The Ecocriticism Reader* (Glotfelty and Fromm 1996), this list of authors and texts was soon augmented to address diverse genres—poetry, creative nonfiction, drama, film—and authors of diverse genders, sexualities, racial and cultural perspectives, as described in *Ecofeminist Literary Criticism* (Gaard and Murphy 1998), *The Environmental Justice Reader* (Adamson, Evans, and Stein 2002), *Black Nature* (Dungy 2009), and *Black on Earth* (Ruffin 2010). The central questions of the new field also developed: "Who and what counts as 'nature' in this text?" ecocritics asked, and "How have diverse cultural traditions conceived of human–environment relations in conjunction with human–human relations?" Environmental justice ecocritics asked: "What standpoint authorizes the circulation of cultural and material toxins through environments and bodies seen as 'nature'?," while African-American ecocritics asked: "Why were Indigenous and African-American texts overlooked in the formation of the American ecoliterary canon?" With the question "What isn't a text?" ecocritics came to recognize the stories immanent within place, within cultural artifacts and environments. By 2014, the vibrancy and agency of matter had gained intellectual currency, prompting the formation of a new subfield, Material Ecocriticism (Iovino and Oppermann 2014).

During the pivotal decade of the 1990s, transgender, queer, and animal studies scholarship also developed in ways that would provide fertile ground for transecology. Kate Bornstein's *Gender Outlaw* (1996) and *My Gender Workbook* (1997) took up where Feinberg left off, narrating the feelings, culture,, and technologies that prompted Bornstein to transition, and to write books that would help allies understand the ways that gender binaries had rigidly constructed even their own gendered identities. Queer ecofeminists explored environmental politics and the intersections among colonialism, heterosexism, and the assault on indigenous humans, animals, and environments (Gaard 1997; Sandilands 2001). Preceded by ecofeminist theorists whose work had already queered the boundaries of gender, species, race, and environments for over three decades (Gaard 2012), Animal Studies emerged as a field in 2007, prompted by Derrida's famous essay "The Animal That Therefore I Am" (2002), and launching the diverse fields of posthumanism, human–animal studies, and critical animal studies. In *Queer Ecologies* (Mortimer-Sandilands and Erickson 2010), ecocritics asked more new questions: In what ways do "understandings of nature inform discourses of sexuality," and in what ways do "understandings of sex inform discourses of nature" (2)? Moreover,

> what does it mean that ideas, spaces, and practices designated as "nature" are often so vigorously defended against queers in a society in which that

> very nature is increasingly degraded and exploited? What do queer interrogations of science, politics, and desire then offer to environmental understanding? And how might a clearer attention to issues of nature and environment—as discourse, as space, as ideal, as practice, as relationship, as potential—inform and enrich queer theory, lgbtq politics, and research into sexuality and society?
>
> (Mortimer-Sandilands and Erickson 2010, 5)

Queer Ecologies featured essays reviewing Bruce Bagemihl (1999) and Joan Roughgarden's (2004) work detailing the variety of same-sex sexual behaviors across the species, from spiders and octopi to seagulls and ducks, and (of course) bonobos; it explored ways in which queer sexual behaviors had produced queer environments, whether neighborhoods or public baths or municipal parks; it described the public homophobia present in the response to the same-sex penguins who hatched and parented an offspring in New York's Central Zoo.

To date, the queer ecocritical focus on sexualities has not captured the critique of heteronormative gender that trans* perspectives address, though the methodologies of ecocritical, queer, and trans* perspectives are influenced by feminist roots. From the start, feminist methodology has been transdisciplinary, centering on the problems of oppression, and working across the disciplines to uncover resolutions promoting justice and inclusion (Stanley and Wise 1983; Harding 1987; Stanley 1990). Like feminism, as Eve Kosofsky Sedgwick has explained, "the word 'queer' itself means across—it comes from the Indo-European root—twerkw, which also yields the German quer (transverse), Latin torquere (to twist), English athwart" (1993, xii). Bridging queer studies and posthumanisms in their Introduction to *GLQ*'s issue on "Queer Inhumanisms", Dana Luciano and Mel Y. Chen (2015, 189) explain:

> the encounter with the inhuman expands the term queer past its conventional resonance as a container for human sexual nonnormativities, forcing us to ask, once again, what "sex" and "gender" might look like apart from the anthropocentric forms with which we have become perhaps too familiar.

Further augmenting Sedgwick's explication of "queer," Eva Hayward and Jami Weinstein in *Transgender Studies Quarterly* (2015) propose "trans*" and specifically "tranimalities" as terms that utilize humanism's exclusion of transgendered and more-than-human animals from consideration, and "enmesh trans* and animals in a generative (if also corrosive) tension leading to alternate ways of envisioning futures of embodiment, aesthetics, biopolitics, climates, and ethics" (201). Just as Monique Wittig (1978) grounded lesbian feminism in her claim that " 'woman' has meaning only in heterosexual systems of thought and heterosexual economic systems," and Donna Haraway launched posthumanism with her claim that "we have never been human" (2008, 1), Susan Stryker (1994) articulated a transgender standpoint by identifying with the figure of Frankenstein's monster, thereby articulating her coming-out experiences as transsexual in a heteronormative culture, and "claiming the transformative power of a return from abjection" (Stryker 2004, 213). These precedents suggest that transecologies will explore gender and gendered environments by working within, across, into, and through the analytical frameworks of gender, race, sexuality, species, and ability, addressing intersections of transphobia, erotophobia, ecophobia, hetero- and homonormativity, humanism, and more.

What aims and outcomes will transecology articulate? At the end of their Introduction, Mortimer-Sandilands and Erickson (2010, 39) conclude that Queer Ecologies point to:

> an ecology that embraces deviation and strangeness as a necessary part of biophilia, sexual pleasure and transgression as foundational to environmental ethics and politics, and resistance to heteronormativity as part and parcel of ecological science and green strategy alike.

Their words provide a fitting prelude to a text with rich potential for transecological critique: Beth Stephens and Annie Sprinkle's ecosexual activist documentary *Goodbye Gauley Mountain* (2013), and their transdisciplinary theory of sexecology (2016).

Joining rural and working-class mountaintop removal activists with queer ecosexuals, *Goodbye Gauley Mountain* brings the tools of queer feminist eco-activism to the mountains of West Virginia and beyond. "We shift the metaphor from earth as mother to earth as lover," says Annie Sprinkle, "to entice people to have more love of the planet." Stephens and Sprinkle define ecosexuality as "a way to create a more connected relationship" with earth. "We like to have skygasms," says Sprinkle in *Goodbye Gauley Mountain.* "Beth and I have intercourse with the air that we breathe." Their Ecosex Manifesto[2] elaborates:

> We shamelessly hug trees, massage the earth with our feet, and talk erotically to plants. We are skinny dippers, sun worshippers, and star gazers. We caress rocks, are pleasured by waterfalls, and admire Earth's curves often. We make love with the Earth through our senses. We celebrate our E-spots. We are very dirty. ... We are everywhere. We are polymorphous and pollen-amorous.

Bringing queer and trans performance artists to West Virginia's embattled mining communities, Stephens and Sprinkle resist the assimilationist goals of a gay and lesbian politics that can be satisfied with humanist marriage equality: instead, they invite local activists to join them in performing ecosexual, trans-species, polyamorous weddings that bridge the urban/rural, queer/straight schisms by affirming a shared and long-standing love of the mountains, celebrating that love through a genderqueer and posthumanist commitment to community and land, flora and fauna.

Transecology inherits this activist and intellectual genealogy.

Notes

1 The deadliest year on record. The Trans Murder Monitoring project (http://tgeu.org/tmm/) collects data worldwide. Despite advances in human rights, transphobic violence continues: it did not end in 1966 with the Compton's Cafeteria Riot, or in 1969 with Stonewall, or even in 1993 with the publication of *Stone Butch Blues*.
2 See http://sexecology.org/research-writing/ecosex-manifesto/.

References

Adamson, Joni, Mei-Mei Evans, and Rachel Stein, eds. 2002. *The Environmental Justice Reader: Politics, Poetics, and Pedagogy*. Tucson, AZ: University of Arizona Press.

Bagemihl, Bruce. 1999. *Biological Exuberance: Animal Homosexuality and Natural Diversity*. New York: St. Martin's Press.

Bornstein, Kate. 1996. *Gender Outlaw: On Men, Women, and the Rest of Us*. New York: Routledge.

Bornstein, Kate. 1997. *My Gender Workbook: How to Become a Real Man, a Real Woman, the Real You, or Something Else Entirely*. New York: Routledge.

Derrida, Jacques. 2002. "The Animal that Therefore I Am (More to Follow)." *Critical Inquiry*, 28(2): 369–418.

Dungy, Camille T., ed. 2009. *Black Nature: Four Centuries of African American Nature Poetry*. Athens, GA: University of Georgia Press.

Feinberg, Leslie. 1993. *Stone Butch Blues*. New York: Firebrand Press.

Feinberg, Leslie. 1997. *Transgender Warriors: Making History from Joan of Arc to Dennis Rodman*. Boston, MA: Beacon Press.

Gaard, Greta. 1997. "Toward a Queer Ecofeminism." *Hypatia* 12(1): 114–37.

Gaard, Greta. 2012. "Speaking of Animal Bodies." *Hypatia* 27(3): 520–26.

Gaard, Greta and Patrick D. Murphy, eds. 1998. *Ecofeminist Literary Criticism*. Urbana-Champaign, IL: University of Illinois Press.

Glotfelty, Cheryll and Harold Fromm, eds. 1996. *The Ecocriticism Reader: Landmarks in Literary Ecology*. Athens, GA: University of Georgia Press.

Haraway, Donna. 2008. *When Species Meet*. Minneapolis: University of Minnesota Press.

Harding, Sandra, ed. 1987. *Feminism and Methodology*. Bloomington: Indiana University Press.

Hayward, Eva and Jami Weinstein. 2015. "Introduction: Tranimalities in the Age of Trans* Life." *TSQ: Transgender Studies Quarterly* 2(2): 195–208.

Iovino, Serenella and Serpil Oppermann, eds. 2014. *Material Ecocriticism*. Bloomington: Indiana University Press.

Kolodny, Annette. 1975. *The Lay of The Land*. Chapel Hill, NC: University of North Carolina Press.

Luciano, Dana and Mel Y. Chen. 2015. "Introduction: Has the Queer Ever Been Human?" *GLQ: A Journal of Lesbian and Gay Studies* 21(203): 183–207.

Mortimer-Sandilands, Catriona and Bruce Erickson, eds. 2010. *Queer Ecologies: Sex, Nature, Politics, Desire*. Bloomington: Indiana University Press.

Roughgarden, Joan. 2004. *Evolution's Rainbow: Diversity, Gender, and Sexuality in Nature and People*. Berkeley: University of California Press.

Ruffin, Kimberly N. 2010. *Black on Earth: African American Ecoliterary Traditions*. Athens, GA: University of Georgia Press.

Sandilands, Cate. 2001. "Desiring Nature, Queering Ethics: Adventures in Erotogenic Environments." *Environmental Ethics* 23: 169–88.
Sedgwick, Eve Kosofsky. 1993. *Tendencies*. Durham, NC: Duke University Press.
Stanley, Liz, ed. 1990. *Feminist Praxis*. New York: Routledge.
Stanley, Liz, and Sue Wise. 1983. *Breaking Out: Feminist Consciousness and Feminist Research*. Boston, MA: Routledge & Kegan Paul.
Stephens, Beth and Annie Sprinkle. 2016. "Ecosexuality." In *Gender:Nature*, Edited by Renee C. Hoogland. New York: Macmillan/Macmillan Interdisciplinary Handbooks.
Stephens, Beth and Annie Sprinkle. 2013. *Goodbye Gauley Mountain*. Fecund Arts. Film.
Stryker, Susan. 1994. "My Words to Victor Frankenstein Above the Village of Chamounix: Performing Transgender Rage." *GLQ: A Journal of Lesbian and Gay Studies* 1: 237–54.
Stryker, Susan. 2004. "Transgender Studies: Queer Theory's Evil Twin." *GLQ: A Journal of Lesbian and Gay Studies* 10(2): 212–15.
Wittig, Monique. (1978) 2002. *The Straight Mind, and Other Essays*. Boston, MA: Beacon Press.

Introduction

Transecology—(re)claiming the natural, belonging, intimacy, and impurity

Anna Bedford

Gender and nature are inextricably materially and conceptually bound. Yet, claiming "nature" and the "natural" is both radical and dangerous for those who occupy trans identities. Transgender people have been frequently and historically constructed as and attacked for being "unnatural," and so for them to (re) claim nature is a radical act. It's also a turn from the metropoles of society to a theoretical place where cisheteropatriarchy might not reign supreme. Imagination of trans identities outside societal confines and oppression is reflected in the language of trans artists and theorists. In 1995, Jeannette Jones published her photography of drag queens and transgender artists in a compilation she titled *Walk on the Wild Side* (Jones 1995), a name evoking both untamed nature and danger, as "Outlaw" does for Kate Bornstein, with its connotations of being dangerous, outside society, at-large, fugitive, and transgressing (1994). Outside the city limits is the "Undomesticated Ground" that Alaimo and others have sought to identify as feminist space (Alaimo 2000). What *Trans* identities and theories add to this is the insistence of boundary crossing, the impossibility of "beyond," the permeability of walls.

These authors seek to insist upon humans as part of nature and ecology; as part of natural spaces as well as socially constructed ones (the two not being mutually exclusive), their narratives focus on in-between spaces such as camps, and fields. Yet, as we foreground ourselves as animals, as part of nature, we must also remember that such associations have been used to the detriment of women, differently gendered Others, the non-white, poor, and disabled, who have been disqualified, targeted, and discriminated against based on their bodies, their materiality, corporeality, their imagined closer proximity to nature and the wild.[1] Association with nature has also been an association with all the lower ranked identifiers on the devalued side of carefully constructed binaries man woman; culture nature; mind body, and so forth. Importantly, Susan Stryker notes that if one ceases to elevate the human over nature then it means giving up privilege:

> It hurts, and is dangerous, to be dehierarchized, to lose human status by falling outside of norms and thereby being subjected to violence, but decen tering the tangled webs of trans-huManinimality nevertheless offers a better

> ethical starting place for enacting our relationship to Being than trying to prop up a spurious anthropocentric privilege. This is where transecology always begins for me.
>
> (Stryker, Foreword, p. xviii, this volume)

The deconstruction of the hierarchies involved in maintaining nature as other is dangerous, particularly for those historically tied to it, but it is part of an ethical move, and also necessarily part of a trans movement that contests binaries and othering.

The exploitation of Others through the maintenance of hierarchies means that trans and ecological Others are intertwined in their subjugation within the cisheteropatriarchy. We can see their interconnection demonstrated when periods of intensified attacks upon trans rights are concomitant with a similar targeting of others in the LGBTQIA community, women, racial minorities, and the environment. For example, during the dominance of the Republican Party over all branches of the US legislature in 2017, political and social attacks on trans rights intensified concomitantly with attacks on the environment. One month after taking office in 2017, President Trump rolled back protections for trans students that had ensured their right to use school bathrooms that matched their gender identity.[2] A conservative swing and increasing transphobia informed a fear of the Other and perception of bathrooms as "contact zones," as the kinds of potential places of hybridity and mixing discussed by several authors in this collection.[3] As Sheila Cavanagh has explained in her insightful and fascinating study of hygiene, abjection, and the gender segregation of toilets:

> We cleanse the boundaries between the masculine and feminine (or separate the two) in public lavatories so as to police the borderland or indeterminate space between these two discursive and material positions. Gender purity is disciplined by hygienic imaginations and rendered sacred, while gender impurity—signified by a discord between gender identity and the way the sex of the body is intercepted by others—is profane.
>
> (Cavanagh 2013, 427)

"Contact zones" and "borderlands" are places of perceived danger, and impurity, but they must also offer liberation and have the potential to be reclaimed as sites of belonging for those cast as "impure" within the social body.

Six months after revoking protections for trans students, President Trump announced that trans service members would be banned from the military. The broad-sweeping proposed ban went against recent trends under the previous administration to open the military to more citizens who want to serve, for example, openly gay and lesbian members with the repeal of DADT[4] in 2011 and the opening of all combat positions to women in 2016. On August 25, 2017, Trump went a step further and ordered the Pentagon not to recruit openly transgender individuals moving forward (Phillip 2017).[5] The military is a "zone" that conjures up images of camaraderie, *close contact*, and unity in the face of an "enemy," all of which have been invoked to exclude numerous "Others"–racial others, women, and those with different sexual orientations.

The military is an important repository for cultural US self-identity at a national level: it is intertwined with narratives of patriotism, strength, and protecting the (feminized) homeland from the Other, or, in Trump's terms, "enemy," which is not just physically outside borders but also culturally. Militarism is tied to conquests and occupation of land, women, and Othered people, as well as defense of the feminized "homeland" or "motherland." The military guards against enemies who would occupy or threaten American soil, but also the "American way of life," which is idealized in popular discourses as well as immigration policies[6] and political narratives as heteropatriarchal domesticity. The most famous trans member of the US military, Chelsea Manning, who transitioned in military prison, is a figure that distilled national fears about threats to the American way of life through her intertwined image among the general public as trans and as national traitor—a dual threat to the nation. The military, as the protector of a nation, remains a site of conservative and frequently toxic cis- and hetero-masculinity, as evinced by the policing of its membership, legally and politically, but also culturally and violently (evinced by widespread sexual assaults and rape), in a climate of sexism, homophobia, and transphobia.

At the same time that trans rights were threatened in the United States—at a federal level—during the Republican swing to power in 2017, environmental ones were too. Trump issued a presidential memorandum to advance construction of the Dakota Access Pipeline that had been halted because of tribal concerns; the "Modernization of the Endangered Species Act" that was introduced to remove what proponents call "regulatory burdens"; a man who did not believe that human activity contributes to climate change was appointed as head of the Environmental Protection Agency (Scott Pruitt was confirmed by the United States Senate as Administrator of the Environmental Protection Agency on February 17, 2017); references to climate change were removed from the White House website; there was initially a plan to remove the climate change page from the EPA, and the list goes on. One response came from the National Park Service employees who created a "rogue" twitter account and began tweeting climate change data. An online embrace of the Park Service by citizens included messages and memes with statements such as "I was not expecting the Park Rangers to lead the resistance." However, perhaps we shouldn't be surprised that environmental advocates stepped to the front of protests in a moment of surging cishereopatriarchy. The similar treatment of LGBTQIA and the environment is not merely coincidental.

To understand some of the links here—why women's, LGBTQIA rights, and environmental ones are simultaneously targeted—one must look at their interconnections and the way in which they are similarly constructed by cisheteropatriarchal worldviews. The colonizing attitude towards women and land is one that frames them as something white men are entitled to—something they can occupy, build pipes through, or indeed "grab by the pussy," should they have one. Women and land are something white men can imagine owning and profiting from. It is perhaps less of a disdain for land that causes its abuse within a capitalist cisheteropatriarchal society and more an intent to profit from it, without concern for the consequences. Men's desire to build, to remove cheap energy, to erode not only natural spaces but also any legal protections for them are all part of an idealized unfettered capitalism and an individualistic masculinity upon which nothing encroaches. This capitalistic, individualistic, cisheteropatriarchal worldview doesn't conceive trans people as property to be profited from, but it does imagine them as a threat to masculinity. Hegemonic masculinity involves a privileged male self and one who finds itself threatened by women in non-traditional roles, by non-heterosexual relationships, and certainly by "*the movement across a socially imposed boundary*" that Stryker defines as "trans" (2008, 1).

The collection opens with Elizabeth Parker's "'The bog is in me': Transecology and *The Danish Girl*," which is an insightful reading of David Ebershoff's novel *The Danish Girl* (2000) and Tom Hooper's film adaptation of the same name. These texts share a fictionalized story of Lili Elbe, one of the first trans women to undergo gender-confirmation surgery. Parker's reading demonstrates how Lili's journey and the novel and film are indelibly linked to and mediated through connections with the natural world, and, in particular, the motif of the Danish bogs that Lili repeatedly paints.

Central to Parker's argument is the refutation of transgender identity as "unnatural." Through interviews with the novel's author, Ebershoff, along with her own close reading of the text, Parker outlines how Lili's identification with nature in the novel establishes her as a "natural" being, thus firmly undercutting the traditionally intolerant views of being transgender as "monstrous" and "outside the natural order" (Chapter 1, p. 24, this volume). Susan Stryker has defined "trans" as "*the movement across a socially imposed boundary away from an unchosen starting place*—rather than any particular destination or mode of transition" (Stryker 2008, 1). The "natural-ness" of trans movement is a recurring assertion in the disparate essays in this volume, one that is made in order to "depathologize" the trans body and transitioning itself, to which we will return.

In "Coming out, camping out: *Transparent*'s eco-ethical approach to gender," Katherine Thorsteinson and Hee-Jung Serenity Joo continue to interrogate the relationship between nature and trans identity through their analysis of the fictional character of Maura from a popular television series. Thorsteinson and Joo begin by considering the spaces of various camps that are constructed as retreats from patriarchal society and city; the ongoing risk of essentializing nature; and how these "nature" retreats exclude as well as welcome.

Crucially, Thorsteinson and Joo explain how trans women are excluded from natural spaces, even those constructed to be feminist and/or queer. The trans character, Maura, finds herself unwelcome at the lesbian feminist music festival, with its "womyn-born-womyn" policy, as well as the "Camp Camellia" retreat for cross-dressers that prohibits hormones. These are camps where women, lesbians, and cross-dressers make space for themselves and retreat for self-reflection, community, and affirmation, all mediated through the natural world. Their various cries of the warning "Man on land!" at Idyllwild Womyn's Music Festival and the rallying "We are cross-dressers, but we are still men!" at Camp Camellia exemplify how trans women are excluded from women-only spaces as well as queer male ones. These exclusions are, as Thorsteinson and Joo note, based in historic realities of gatekeeping and trans rejection from similar non-fictional camps (p. 32).

Thorsteinson and Joo show how, even as others retreat from mainstream society to find themselves with nature, trans women like Maura are unable to find their place, and, in fact, face alternative systems and measures of authenticity and purity. The authors argue that "the show reveals how ironic camp aesthetics can do just as much to reinscribe as to subvert gender essentialism" (p. 32). They carefully link ideas of purity to other kinds of camps, including Nazi concentration camps, and a history of eugenics (p. 32). Thus, this chapter presents an astute analysis of one trans character's interaction with nature "camps" as a means of a broader examination of possibilities for escaping the hierarchies of dualisms, dualisms themselves, and engaging with nature and other people in more empowering and *intimate* ways; ways that make space for trans identities.

In "Posthuman ecological intimacy, waste, and the trans body in *Nånting måste gå sönder* (2014)," Wibke Straube continues Thorsteinson and Joo's suggestion of intimacy as a central strategy for transecology. Straube analyzes the intimacies between gender nonconforming human bodies and the nonhuman in a film by the Swedish trans filmmaker, Ester Martin Bergsmark. Straube suggests *posthuman intimacy* as a way beyond dualisms and works to break down any "purity" of the natural that Thorsteinson and Joo have demonstrated as so problematic by carefully highlighting pollution and waste within Bergsmark's depictions of the "natural." Waste is subjective and itself a cultural construct, argues Straube (pp. 65–66), and, along with impurity, is a part of LGBT history (pp. 66–67).

Ideas of purity have been an essential mechanism to control women (for example, the nineteenth-century ideology of "true" womanhood and cult of domesticity that were tied to "nature"), to put both women and nature upon pedestals, and also used to exclude individuals and behavior that deviates from binary constructions of masculinity and femininity or male and female (for example, as Thorsteinson and Joo show Maura confronted with in the "womyn-born-womyn" community and a purity of "nature" that Straube deconstructs). Although several of these essays cite various examples of trans women's exclusion from women-only spaces, trans men are similarly excluded from masculine spaces. For example, only in 2017 did the Boy Scouts of America—historically linked to masculinity and nature, and the construction of masculinity and brotherhood *through* interactions with nature—begin accepting trans boys, a move that came only months before announcing they would accept girls. Straube furthers the deconstruction of purity that is also central to other authors in this collection, arguing that humans are *more* than human, containing other species (e.g., bacteria). They suggest, after Alaimo, that trans-corporeal movement is a "rebellion against boundary politics" and a "significant … 'contact zone' between human and more-than-human nature" (p. 67).

Straube's "contact zones" reintroduce questions of space. Thorsteinson and Joo look at a couple of nature retreats in their study of camps in "Coming out, camping out," and carefully illustrate the flawed human attempts at retreat and connection within such spaces, which are necessarily imagined by campers as a contrast to the city and mainstream society. Straube offers a highly compatible analysis of natural spaces that are impure and in contact—from the urban wilderness of malls to the trash that floats through the nature scenes within the film they focus upon. The suburbs, suggests Straube, offer a home outside city regulation—these are spaces for "the queers, the poor, the homeless, and garbage itself" and the site of "refugees' temporary campsites, homeless people's tents, [and] cruising areas" (p. 63). Straube's suburbs invoke for me a modern and transatlantic interpretation of Shakespearean Liberties, and, like the Liberties, just outside the city walls, the suburbs show the limits of heteropatriarchal power and indeed anthro-centric power. The Jew of Malta was thrown over the city walls to be devoured by birds and beasts, and still, the outcast, polluted, or impure bodies are thrown from the city, and, as Straube's chapter shows, cinema itself is also of the suburbs, as the playhouses were of the Liberties. The contact zones described by comparativist scholar Mary Louise Pratt (1992) are places of what anthropologist Fernando Ortiz (1995) calls "transculturation" and these are places of trans identity and art that eschew "crossing" and insist instead upon impurity and mixing.

Julia Kuznetski's "A journey through eco-apocalypse and gender transformations" discusses Angela Carter's 1978 dystopian novel, *The Passion of New Eve*. It's a novel in which a self-centered and chauvinistic young man, Evelyn, flees the city to find himself in the wasteland of desert, which appears to him as a kind of postmodern pastoral space (p. 84). Ironically, it is a place inhabited by lizards, birds, and an entire underground enclave of militant feminists who kidnap Evelyn, transform his body, and artificially inseminate him. Kuznetski illustrates how Eve/lyn and the character of Tristessa—a beautiful actress and a public exemplar of femininity who is revealed to be a trans woman and, later in the novel, identifies as a man—together center transgression, confusion, crossing, and ambiguity (pp. 92–93), and a mixing of substances foreshadowed by the Czech alchemist in the opening chapter of the novel.

Kuznetski, like Straube, focuses her analysis on the transgressive and transcorporeal. Discussions of purity and (in)authenticity (p. 82, p. 84, pp. 89–90) again resurface here, and reverberate from Straube, as well as Thorsteinson and Joo's writings. In particular, Kuznetski describes how the character of Tristessa, the movie star who was Evelyn's childhood fantasy, exposes the "contingency of gender" that Judith Butler (1999) insists upon (p. 000), and the performativity of ideal womanhood.

Various authors in this collection have strong and opposing feelings about Judith Butler. Some reject her while others embrace her. I believe, as Butler asserts, one can claim that gender identity is ontological while also asserting that gender is performative. One is not born a woman but becomes one, as Beauvoir (2001) says. Sex assigned at birth is not what makes us masculine or feminine, men or women, or neither. I believe that gender is performative, a process of becoming over and over through a million interactions each day. My own trans students have frequently rebelled against Butler at their first reading. "My gender is not a performance" says Julia Serano (2010), and this is often their cry too. However, later, many students grow to appreciate Butler and come to learn from her that their gender is as real as anyone else's; that gender can be performative without being a mere performance. Gender is absolutely real—for trans people too often it is a matter of life and death—and it is an essential part of the way we interact with the world. Our bodies are often the "contact zones" that Straube analyzes and the "ecotones" Anae writes of in Chapter 8.

In Kuznetski's reading, nature is foregrounded through an imminent eco-apocalypse, through the landscape of the desert, and through the treatment of trans bodies as landscapes. Kuznetski (p. 81) writes:

> Tristessa's self-perception as a woman in a man's body is recounted as a kind of journey: "For hours, for days, she had wondered endlessly within herself, but never met anybody, nobody." Thus, body, identity, and landscape are linked in *Eve*, although we cannot speak of an "end" to the journey for the protagonist, nor a "true" gender arrived at as a result.

The failure to finally *arrive* at a " 'true' gender" returns us to the impossibility of purity, or dichotomous gender. In turn, this failure to be "properly gendered," argues Kuznetski, creates what Butler called "abjects," who in failing (or escaping?) "proper" gendering consequently fail to fit into the category of "human" (p. 92). This in itself, asserts Kuznetski, opens up possibilities for greater "interrelatedness and connection" with nonhuman life (p. 92), or, as Straube put it, *posthuman intimacy.*

In "Chinese literature, ecofeminism, and transgender studies," Peter I-min Huang discusses a prominent Taiwanese woman poet and the challenges her writing poses to mainstream nature/culture dualisms. By the late 1990s, writes Huang, Taiwanese women had gained considerable political and economic power (p. 97), and, he argues, Chen is among a class of "avant-garde poets who liberated literature in Taiwan from the patriarchal confines imposed upon it during the period of martial law and earlier" (p. 98). Chen's "in-between" poems ("In-between" being the title of her anthology) resist dualisms and, claims Huang, "powerfully represents in-between states of identity and erotic modes of being that resonate with both transgender studies and ecofeminist theory" (p. 98).

Huang points to one example of Chen's intervention in the treatment of women by patriarchal culture and literature by considering the dismissal of non-reproductive women, and the ways in which Chen recuperates them beyond biological imperatives and "survival of the fittest" (p. 101). This is important because such women resist cisnormative and heteropatriarchal roles and have consequently been vilified as "unnatural," and as not fully women. Resistance to their vilification and dismissal is also important because reproductive capabilities have, at times, been central to attacks upon the LGBTQIA community, as well as key to non-consensual assignment of sex at birth for intersex individuals. In addition, women who don't reproduce are linked to queer women and considered dangerous because their sexuality is not contained as society demands, and their energies not directed into the—often all-consuming—caregiving work of mothering. This is to say that women who do not reproduce offer a clear threat to patriarchal order not only through their failure to *re*produce but also the threat of their potential *production* (when their energy is not consumed elsewhere).

In his analysis, Huang also points to the traditional and recurring speciesist vilification of women through their association with nature, and in particular with figures of snakes (pp. 101–102). Huang notes that snake imagery associated with women has a long literary history in both the East and West, along with connotations of both as poisonous and treacherous. Indeed, one might look at one of the most well-known affiliations of women and snakes—that of Eve and the snake in the Garden of Eden—as an association that brings about the downfall of humankind. The snake-like threat of non-reproductive women, then, is a threat to others, and women who are not appropriately contained within hetero- and reproductive marriages and roles are similarly a threat to the social order. As a treacherous reptile, such women are a *natural* threat to culture and social orders. Yet, as Huang goes on to note, these dichotomies between human and nonhuman animals as either part of nature or apart from it will not hold up, and he illustrates, after Haraway and Alaimo, that animals are also "cultural beings" (p. 102).

Huang closes by examining the treatment of a Chinese God/dess, Guanyin by the famous sixteenth-century Chinese writer, Cheng'en Wu. Guanyin, argues Huang, is an Eastern trans deity who traverses genders and human-nonhuman animal boundaries (p. 106). Thus, Huang asserts that Chen and Wu both offer in-between characters in terms of gender and species, and he responds to Stryker's assertion that European culture is "haunted" by "transgender phenomena" by adding his own analysis of trans phenomena in key Taiwanese and Chinese literature and religion (p. 107).

In Chapter 6 of this collection, Mat Fournier offers a trans reading of Annemarie Schwarzenbach's writing, in which he identifies recurring themes of queer love and gender disruption (p. 114). Schwarzenbach is known as a cisgender woman and lesbian author, and thus it is important to note that it is Fournier's *reading* itself rather than the author or subject, as in other chapters of this collection, which constitutes the *trans* turn and perspective. Fournier explains, "my aim is to shift the perspective from sexual orientation to gender disruption" (p. 111).

Fournier opens with an examination of Schwarzenbach's life, emphasizing how her homosexuality and her masculine womanhood disrupted gender norms of the time. He points, in particular, to her androgyny. Although Fournier distinguishes her androgyny from masculinity, more often than not androgyny requires a casting off of all things feminine, not a mixing of feminine and masculine dress, manners, or attributes. Androgyny can present itself as neutral in the way that universal pronouns and words do, which is to disavow the feminine, which is always particular. Fournier also notes that Schwarzenbach is repeatedly referred to and associated with the angelic, and "angels, as we know, have no sex" (p. 114). This is not entirely convincing, since in keeping with the sexless or neutral being defined as simply not-female we see that the representations of angels in the Bible are male forms. Thus, perhaps one can argue that, for women, embracing androgyny is not gender neutrality (I maintain there is no such thing) but clearly gender disruption. In addition, Fournier also notes that Schwarzenbach was an unmarried woman, with all the accompanying eschewing of social categorization and challenge to heteropatriarchal order—so often asserted as "natural"—discussed above in relation to Peter I-min Huang's chapter.

Fournier then turns to the landscapes and representations of nature in Schwarzenbach's writing and argues that "the Orient becomes another word for disorientation" (p. 118), another way in which Schwarzenbach's narratives are disruptive. The environments themselves are "inhospitable" (p. 112) and threatening to bodies. It is important to note that the *Orient* (meaning *East*) in English is used to denote Asia, but, in this chapter the author uses Orient in the Germanic manner, to include Arabic-speaking countries and Persia.

In "Transplacement: nature and place in Carter Sickels's 'Saving' and 'Bittersweet,'" Katie Hogan calls our attention once more to rural spaces and to the ways in which these places and communities have been depicted as *hostile* environments to queer and trans people in both popular narratives and LGBTQIA literature. She offers an astute analysis of home, belonging, and not-belonging for the rural Appalachian-born characters in the work of trans author Carter Sickels. In particular, Hogan highlights the perception and representations of violence that is disproportionately displaced from urban to rural areas, along with concomitant depictions of the characters that live rurally.

Given the cultural imagination of the countryside as inhospitable to the queer and trans, Hogan argues, after Kath Weston, that rural communities are generally cast as dangerous closets from which queer people must "get thee to a city" (p. 130). Carter Sickels' trans characters, however, find ways to return, as Hogan illustrates. Sickels, she argues, manages to capture the multiplicity of the country rather than the flat caricatures of "backwardness" or closed mindedness. These are places with networks of belonging that include and exclude in line with values and affinities, of which queerness is only one identity claim (pp. 132 133). Along with strong senses of community, family, and neighbor hood, Hogan notes that Sickels also shows the complicated influences of religion, poverty, drug abuse, and environmental degradation that also make up these multi-dimensional urban communities, as well as a beauty of "home," inescapable even for Sickels' characters, like Dean, who fled to a city. Of Sickels' work, Hogan (p. 139) writes:

> Instead of portraying urban space as The Promised Land and rural areas as The Wasteland, he, like many rural queer studies scholars, offers a more

> capacious and complex vision, one that neither eclipses nor exaggerates a particular space.

Hogan's chapter is thus an important meeting of rural and queer studies that complicates the way in which the two have been divided and positioned as antagonistic.

Nicole Anae's "Sexuate ecologies and the landmarking of transgender cultural heritage" continues this vital discussion of place and belonging for transgender citizens and their narratives. Anae's is a fascinating analysis of "landmarking" transgender cultural heritage. From the "theme studies" of the National Park Service that enable a "topic" such as LGBTQIA history to be a lens for recognizing historic landmarks, to the interpretations of UNESCO's "cultural heritage," Anae points to an implied audience and a recognizable shared heritage as key to a sense of national belonging (p. 147). In the United States, one is reminded of the very recent designation of Stonewall as a National Monument (and thus part of the National Park System) by President Barack Obama in 2016.

Anae reminds us, powerfully, that landscapes and ecologies are part of cultural heritage. This, of course, belies the idea that the natural, the land, and nature are set apart from culture and the human. Anae writes also of "intangible" heritage and argues convincingly for the consideration of culture as part of a broader ecology. She then turns her attention to "school-scapes" as ideological landscapes and cultural heritage and considers their impact upon trans children in Australia. Within this discussion Anae focuses on "ecotones" as places of transition and possibility. These "ecotones" of "school scapes" constitute an additional kind of "contact zone" to those considered in Straube's chapter in this collection. Anae argues that, ultimately, in its various cultural and ecological forms, "landmarking" is a process of collective valuing and public recognition.

Next, in "Transgender: an expanded view of the ecological self," Gail Grossman Freyne, with a background as a therapist, grapples from a somewhat medicalized perspective with the possibilities opened up by transgender identities in conceptualizing ecological self. The "Ecological Self" of Deep Ecology, as articulated by Arne Naess, rejects individual selfhood in favor of connections between human and nonhuman nature and between humans. Ecofeminists have challenged some of the gendering of an "ecological self," yet it has provided a fruitful model for concepts of connection with human and nonhuman nature, and ecofeminist Val Plumwood (1993, 1995) has advanced an "Ecological Selfhood" that also incorporates feminist practices of *caring*, including caring within scientific models.[7]

"By overlaying our sexual difference with the social construction of gender we are left with two incomplete versions of human nature: the masculine and the feminine," writes Freyne (p. 177). This is, of course, true, but Freyne herself soon begins to hint at the social construction of not only gender but also sex. In the (nonhuman) animal kingdom, she notes, female hyenas have penis-like structures, male fruit bats lactate, males gestate, and so forth, and there is a natural occurrence of intersex bodies (p. 181). Indeed, instances abound, and evolutionary biologist Joan Roughgarden has documented examples of "sequential, simultaneous, and crisscrossing hermaphrodism" in fish, that demonstrate that "male and female functions don't need to be packaged into lifelong distinct bodies" (Roughgarden 2013, 154). Although she doesn't state it in such terms, Freyne's "natural world" examples highlight our manufactured expectations of sex and illustrate how the idea of binary sexes (in addition to binary genders) is a social rather than a "natural" one. Freyne argues that we should center transgender experiences in our visions of ecological selfhood because they illuminate the possibility of "the full range of human behaviors" (p. 177) and can help us understand and recognize difference without hierarchy, as ecofeminists have sought to do with regard to gender and the natural world.

Freyne writes that trans individuals are a "special synthesis of mind and body, of reason and nature, living with a neurological condition rather than a psychological one" (p. 182). It is worth noting that gender theorists such as biologist Anne Fausto-Sterling (2000) have demonstrated how sex is *socially* rather than *biologically* or "naturally" constructed as binary. Thus, since the dualisms are a social construct, not only trans and intersex people but also cisgender people are always simultaneously "male and female" and "masculine and feminine." To be sure, the authors in this anthology celebrate such "impurity," but to suggest that trans individuals are alone in being a blend of masculine and feminine or to argue that trans women or trans men are less women or men than cis women and cis men and instead a "special synthesis" demonstrates a clinical point of view that runs counter to recent gender theory and trans advocacy.

Freyne goes on to problematically mobilize narratives of being "trapped in the wrong body," which have accompanied discussions of trans experience over and over in recent decades. In many ways such narratives appeal to the cisgender community that might be able to imagine how they would feel if their firmly cisgender self were to wake up in the body of the supposedly opposite sex. Popular culture has depicted "wrong" bodies and trapped minds in comedies such as *Freaky Friday* (1976, 2003), *Big* (1988), and many others, and we can even find a mind trapped without a body in *The Man with Two Brains* (1983). But serious consideration of a mind trapped in the wrong body runs up against the difficulty of imagining how an identity might be divorced from a body. It is possible to say that some men have vaginas and some women have penises, and those bodies are not "wrong," just as it is possible for some male bats to lactate and other animals' bodies to defy sex expectations. At the same time, surgeries, if they are chosen, may be vital and necessary but also "confirming" rather than "correcting" or "reassigning." Freyne's chapter, even as it seeks to celebrate trans identities and difference, exemplifies a clinical approach that, unfortunately, has been used all too often to pathologize trans bodies, and provides an example of how approaches and the narratives mobilized by them vary as we move between fields.

In "Good animals: the past, present, and futures of trans ecology," Nicole Seymour considers methods to "depathologize trans embodiment" (p. 192), and just as Freyne, from the clinical perspective, tries to reclaim transitioning as part of nature, Seymour also works in "opposition to [transitioning's] dominant framing in popular and medical discourse as an 'unnatural,' technoscientific intervention" (p. 191). This argument can also be expanded to recognize that nature and technoscience are not at odds, since as Myra Hird has claimed, "At a basic level, life itself is, and has always been, technological in the very real sense that bacteria, protoctists and animals incorporate external structural materials into their bodies (Margulis and Sagan 1997)" (Hird 2013, 162). In "Animal Trans," Hird argues:

> This use of technology to distinguish between nature and culture obscures the very real and energetic invention and use of technology by non-human living organisms (termite high-rise cities include "birth chambers, hatcheries, the insect equivalent of schools, hospitals, honeymood quarters, workshops and morgues" all under sensitive climate control) as well as the extent to which so-called human technologies actually mimic technology already invented by other species (Margulis and Sagan 2002). The continued focus on technology also further limits the discussion to transsex rather than considering the lived experiences of transpeople more generally.
>
> (Hird 2013, 162–3)

Seymour similarly points to nature as a site of transitioning, in opposition to its image as a kind of exclusively human artifice.

Katherine Hayles claimed, in 1999, that we are all "posthuman," and Donna Haraway has long proposed that "we are all chimeras, theorized and fabricated hybrids of machine and organism; in short, we are cyborgs" (Haraway 1991, 150). Further, in his argument for the interdependence of humans and nonhumans, and against the possibility of static identity, Bailey Kier claimed, in 2013, that "everybody on the planet is now encompassed within the category of transgender" (Kier 2013, 189). In the final chapter of this collection, Seymour similarly wonders, in a way, "Are we all trans now?", and what are the important ethical consideration we must balance in such identifications? "Expanding 'trans' threatens to obscure the environmental and other injustices (and other experiences) unique to transgender people," she suggests, "but expanding 'trans' also potentially destigmatizes transgender embodiment by showing it is not a unique experience" (p. 200). In addition, Kier has used the expansion of trans not only to destimatize but also to destabilize, to decenter the human. As she pushes to expand what should be incorporated under the classification of trans, Seymour ponders further ethical questions of human responsibility during a period of anthropogenic climate change and the consequences of decentering ourselves in such a moment.

In *Transecology* we find converging fields, and theories, but also the insistence upon material and bodily experiences. Too often labs, classrooms, and conferences—with their emphasis on the labor of the mind—seek to disavow the corporeal. In *Transecology* we need transgressive insistence upon the bodies of human scientists, academics, and activists, especially to including trans bodies; we need to recognize our own specificity, partiality, *immodesty.* Haraway has demonstrated the impossibility of the "Modest Witness," and the exclusion of women, non-male, non-white, and working-class others from the pursuit and production of knowledge, which is to say the position of subject rather than object within scientific endeavors and discourses (Haraway 1997). Instead we can and must imagine *immodest witnesses* who embrace and emphasize their partiality, physicality, and embodied participation as a more revolutionary and honest approach to facts and science (Bedford 2011, 2014).

This collection offers a wonderful traversing of disciplines and the perspective of thinkers from a variety of fields of expertise, countries, and identities, offering disparate approaches to transecology. It claims the natural; it relishes the impure; it works toward connections, intimacy, and belonging. I believe it will appeal to activists, academics, those in Gender Studies and those working in Environmental Studies, among many others. Its many crossings and intersections and centering of trans approaches will inspire its disparate audiences. May the *movement* continue.

Notes

1 The association of women with nature has also been used to nature's detriment, as many early ecofeminists have cautioned. The efficacy of "Mother Earth" images and discourses, for example, within a patriarchal culture where mothers are not respected is questionable at best. It may, also, simultaneously trivialize and cast as domestic the violence against it, as Stacy Alaimo argues:

> Mother Earth ideology here codes the earth and by feminine association, women, into passive victims at the same time that it depicts polluters as mere naughty boys, thus making the problem personal and familial instead of political and systemic. It shifts the focus from patriarchal capitalism to the home and places the blame and responsibility, not on corporate polluters, scandalous lack of government controls, or waste-oriented capitalism but ultimately on homemakers, who had better use cloth diapers and keep those pots fully covered.
>
> (Alaimo 1994, 137)

2 So-called "bathroom bills" had popped up in the US at state levels to require people to use the bathroom of their assigned sex. Several of these kinds of bills were put forward over preceding years but failed to pass into law until one was enacted by North Carolina in March 2016.

3 The arguments for restricting bathroom use focused on safety, playing upon patriarchal fears, and mobilizing paternalistic concern for young girls in the bathroom with trans women. The result was to make trans people less safe, and more likely to be victims of transphobic violence (which was already at alarming levels) and to restrict

their movement, their access to public spaces, their ability to attend schools successfully, and to participate in public life, because our ability to be present in public spaces is restricted when we are prohibited from using bathrooms.

4 Don't Ask, Don't Tell (DADT), seen as a progressive move in 1994, allowed Americans to join the armed forces regardless of their sexual orientation but demanded their closeting, and it continued a prohibition on openly gay members that argued that their presence was detrimental to morale and order. This argument is echoed in the language of "disruption" suggested by Trump about trans personnel in 2017.

5 The reported rationale was curtailing "disruption," as was argued of other groups in the past, but also "tremendous medical costs." In an era of neoliberalism, amidst ongoing debates about healthcare costs and free market for civilians in the US, the appeal to fiscal conservatism was a popular one. What it did, however, was to paint trans military members as a monolithic group, when in reality one's identity as trans is not synonymous with an ongoing medical condition. Not all who identify as trans will medically transition, and if they do then those transitions take a variety of forms, requiring different medical care at different stages. A brief consideration of costs and alternative actions proves that a ban on trans citizens in the US military was securely rooted in transphobia rather than fiscal responsibility. Ultimately, the motivation of avoiding "tremendous medical costs" is belied by the fact that Trump could have banned coverage for such medical expenses rather than banning trans members themselves (though such a move would be unethical and exclusionary). Coverage for gender-confirmation surgeries of active duty military members was announced by Pentagon officials under the Obama administration in 2016, just one year earlier, yet Trump planned to oust trans members from the military rather than roll back medical coverage to curb "tremendous" costs. The fiscal rationale is further exposed as a convenient rouse or red herring when one considers the costs involved, which a 2016 study by the Rand Corporation estimated to be between $2.4 million and $8.4 million annually. This figure is put into perspective by the *Washington Post* and the BBC, among others, which reported days after Trump's announcement that the "tremendous" costs Trump targeted were dwarfed many times over by the expense of covering erectile dysfunction medication for the military (Ingraham 2017, BBC News 2017).

6 For a thorough study of the history of US immigration policy and its policing of patriarchal norms and heterosexuality see Eithne Luibheid's *Entry Denied: Controlling Sexuality at the Border* (2015).

7 See Kheel (1991), Spretnak (1997), and Zimmerman (1990) for an in-depth analysis of Ecofeminism and Deep Ecology.

References

Alaimo, Stacy. 1994. "Cyborg and Ecofeminist Interventions: Challenges for an Environmental Feminism." *Feminist Studies* 20(1): 133–52.

Alaimo, Stacy. 2000. *Undomesticated Ground: Recasting Nature as Feminist Space.* New York: Cornell University Press.

BBC News. 2017. "Why Does the US Military Buy So Much Viagra?" *BBC*, July 27. Accessed July 31, 2017. www.bbc.co.uk/news/amp/40741785.

Beauvoir, Simone de. 2001. *The Second Sex* (1949). New York: Random House.

Bedford, Anna. 2011. "Reluctant Travelers: Vonarburg's Postcolonial Posthuman Voyagers." *Femspec* 11(2): 84–104.

Bedford, Anna. 2014. "Speculative Survival: An Ecofeminist Analysis of Canadian Women's Science Fiction." PhD diss., University of Maryland College Park.

Big. 1988. Directed by Penny Marshall. Written by Gary Ross and Anne Spielberg. Performed by Tom Hanks, Elizabeth Perkins, and Robert Loggia.

Bornstein, Kate. 1994. *Gender Outlaw: On Men, Women, and the Rest of Us*. New York: Routledge.

Butler, Judith. 1999. *Gender Trouble: Feminism and the Subversion of Identity*. New York: Routledge.

Cavanagh, Sheila. 2013. "Touching Gender: Abjection and the Hygienic Imagination." In *The Transgender Studies Reader 2*, edited by Susan Stryker and Aren Z. Aizura, 426–42. New York: Routledge.

Fausto-Sterling, Anne. 2000. "Dualing Dualisms." In *Sexing the Body: Gender Politics and the Construction of Sexuality*, by Anne Fausto Sterling, 1–29. New York: Basic Books.

Freaky Friday. 1976. Directed by Gary Nelson. Written by Mary Rodgers. Performed by Barbara Harris, Jodie Foster, and John Astin.

Freaky Friday. 2003. Directed by Mark Waters. Written by Mary Rodgers, Heather Hach, and Leslie Dixon. Performed by Jamie Lee Curtis, Lindsay Lohan, and Mark Harmon.

Haraway, Donna. 1991. "A Cyborg Manifesto: Science, Technology, and Socialist-feminism in the Late Twentieth Century." In *Simians, Cyborgs and Women: The Reinvention of Nature*, 149–81. New York; Routledge.

Haraway, Donna. 1997. *Modest_Witness@Second_Millennium.Femaleman_Meets_Oncomouse: Feminism and Technoscience*. New York: Routledge.

Hayles, N. Katherine. 1999. *How We Became Posthuman: Virtual Bodies in Cybernetics, Literature, and Informatics*. Chicago, IL: University of Chicago Press.

Hird, Myra J. 2013. "Animal Trans." In *The Transgender Studies Reader 2*, edited by Susan Stryker and Aren Z. Aizura, 156–67. New York: Routledge.

Ingraham, Christopher. 2017. "The Military Spends Five Times As Much on Viagra As It Would on Transgender Troops' Medical Care." *The Washington Post*, July 26. Accessed July 31, 2017. www.washingtonpost.com/news/wonk/wp/2017/07/26/the-military-spends-five-times-as-much-on-viagra-as-it-would-on-transgender-troops-medical-care.

Jones, Jeanette. 1995. *Walk on the Wild Side*. New Jersey: Barricade Books.

Kheel, Marti. 1991. "Ecofeminism and Deep Ecology: Reflections on Identity and Difference." In *Covenant for a New Creation: Ethics, Religion, and Public Policy*, edited by Carol S. Robb and Carl J. Casebolt, 142–5. New York: Orbis Books.

Kier, Bailey. 2013. "Interdependent Ecological Transsex: Notes on Re/production, 'Transgender' Fish, and the Management of Populations, Species, and Resources." In *The Transgender Studies Reader 2*, edited by Susan Stryker and Aren Z. Aizura, 189–98. New York: Routledge.

Luibheid, Eithne. 2015. *Entry Denied: Controlling Sexuality at the Border*. Minneapolis: University of Minnesota Press.

Man with Two Brains, The. 1983. Directed by Carl Reiner. Written by Carl Reiner, Steve Martin, and George Gipe. Performed by Steve Martin, Kathleen Turner, and David Warner.

Ortiz, Fernando. 1995. *Cuban Counterpoint: Tobacco and Sugar* (1940). Translated by Harriet de Onís. Durham, NC: Duke University Press.

Phillip, Abby. 2017. "Trump Directs Pentagon to Implement Ban on Transgender Service Members, Bans Sex-reassignment Surgery." *The Washington Post*, August 25.

Accessed August 25, 2017. www.washingtonpost.com/politics/trump-directs-pentagon-to-implement-ban-on-transgender-service-members-bans-sex-reassignment-surgery/2017/08/25/c8c604d8-89e4-11e7-9ce7-9e175d8953fa_story.html?utm_term=.97e9be247150.

Plumwood, Val. 1993. *Feminism and the Mastery of Nature*. London: Routledge.

Plumwood, Val. 1995. "Has Democracy Failed Ecology? An Ecofeminist Perspective." *Environmental Politics* 4(4): 134–68.

Pratt, Mary Louise. 1992. *Imperial Eyes: Travel Writing and Transculturation.* London: Routledge.

Roughgarden, Joan. 2013. "Sex and Diversity, Sex Versus Gender, and Sexed Bodies: Excerpts from *Evolution's Rainbow: Diversity, Gender, and Sexuality in Nature and People.*" In *The Transgender Studies Reader 2*, edited by Susan Stryker and Aren Z. Aizura, 147–55. New York: Routledge.

Serano, Julia. 2010. "Performance Piece." In *Gender Outlaws: The Next Generation*, edited by Kate Bornstein and S. Bear Bergman, 85–8. Berkeley, CA: Seal Press.

Spretnak, Charlene. 1997. "Radical Nonduality in Ecofeminist Philosophy." In *Ecofeminism: Women, Culture, Nature*, edited by Karen J. Warren, 425–36. Bloomington: Indiana University Press.

Stryker, Susan. 2008. *Transgender History*. San Francisco, CA: Seal Press.

Zimmermann, Michael E. 1990. "Deep Ecology and Ecofeminism: The Emerging Dialogue." In *Reweaving the World: The Emergence of Ecofeminism*, edited by Irene Diamond and Gloria Orenstein, 138–54. San Francisco, CA: Sierra Club Books.

1 "The bog is in me"

Transecology and *The Danish Girl*

Elizabeth Parker

David Ebershoff's novel *The Danish Girl* (2000) tells the story of an enchanting, beautiful individual intertwined with Nature. "He was born on a bog," we are told of the text's protagonist, "a little girl born as a boy on the bog" (200).

With recent years, as our relationship as a species to the Earth grows ever more precarious, we have witnessed a huge increase in both scholarship and general interest in the so-called "nonhuman turn." The ostensibly neat and even sacrosanct divisions of "human" and "Nature" are proving to be no longer satisfactory, as our very definitions of each of these terms—especially when construed as binary opposites—seem now perennially in question. More and more we encounter rejections of established thought which views humans as distinct, singular, and closed beings—and find instead overdue and necessary emphases on humans' interconnectedness, plurality, and openness with and to the nonhuman world (see, e.g., Gaard 2010; Giffney and Hird 2008; Anderson et al. 2012). The widening discussion about the dangers and interrelations among neoliberalism, individualism, and environmental crisis has been coupled with a growing recognition of the essentiality of intersectionality and interdisciplinarity. In a time when the co-existence of the human and nonhuman is in many ways catastrophic, it is essential that we continue to expand and nuance this conversation, as we must seek to question and explore—on many different fronts—the ways in which we think about these two constructs.[1] In short, we must *queer* the interrelations between them.

Queer Ecology—a relatively nascent intersectional and interdisciplinary field—is one area in which such research is thriving. It explores the numerous and varied connections that can be made between Queer Studies and Environmental Studies, all the while stressing the importance of "rethinking" (Giffney and Hird 2008, 10) our understandings of the human/nonhuman, as it critiques "normative anthropocentrism" (3). Although the bringing together of Queer and Environmental Studies is in some ways a quite new (and to some surprising and perhaps even contradictory (Azzarello 2008, 138)) innovation, the two in fact have a considerable amount in common. Both, essentially, center on interrogating ideas of "constructedness" and "naturalness"—exploring and deconstructing, for example, such meta terms as "gender" and "Nature"—and together can create a "productive disturbance" (Azzarello 2008, 140). Queer Ecology brings such ideas together, with a specifically "queer" lens, examining their interconnections. Enormous amounts of research exist, of course, on the interrelations between heteronormative, binary genders and the natural world—think, for example, of our associations between "masculinity" and ordered, cultivated Nature (Roberts 2008), and "femininity" with either wild, untamed nature or with sanitized maternal imagery (Scharff 2003)—but there is much less on the potential intersections between *queerness*, in its infinite forms, and Nature. Most relevantly to this collection, despite the wealth of research into "the environment" and "masculinity"/"femininity," there is very little indeed on the potential interconnections between "the environment" and specifically "*trans-*"[2] experiences.

This comparative dearth of work on trans-/environmental connections—on *transecology*—is surprising for several reasons. First, one of the central ideas in the nonhuman turn is that of "trans-corporeality" (Alaimo 2010): a term which of course literally includes "*trans-*" and so is likely to evoke a number of related connotations.[3] Essentially, trans-corporeality refers to the idea that our bodies are porous and perennially intermeshed with the nonhuman, meaning that we are never "separate" from the environment in which we live and are always more-than-human. It firmly underlines interconnectedness. Although Alaimo's use of the prefix "trans-" is not explicitly linked to Queer Studies, the term "trans-corporeality," as Seymour (2017, 255) argues, is "conceptually aligned with transgender studies": it is not only nonbinary but non-*boundary*, always "moving" and "crossing" (2014, 2). Moreover, just as trans-corporeality brings together the human and nonhuman, erasing or at least blurring the boundaries, some have argued that there is a naturally closer relationship between trans-individuals and the nonhuman world. For example, trans-singer Anohni speaks of a "feral, empathic connection with the world," theorizing that it is in the "nature of the transgender person" to have "an increased sensitivity to their environment." Furthermore, it is reasonable to assume that a field entitled Queer Ecology is interested in representing *all* elements of the LGBTQIA+ acronym, including, of course, the "T" for "trans-." Indeed, a further connection that might occur between "trans-" and "ecology" is the fact that both the "trans-" from LGBTQIA+ and "Nature" have a history of being *backgrounded* to more "mainstream" human activity. We must, as Gaard asserts, "tackle ecophobia and erotophobia alike" (2010, 116). Finally, if Queer Ecology—which simultaneously "queers" the nonhuman and "greens" the queer (O'Rourke 2008, 8)—is about challenging fixed ideas and embracing openness, then the very ideas associated with "trans-," which is "multivalent" (Seymour 2014, 2), surely have a rightful place within Queer Ecology.

This chapter seeks to explore some of the interconnections involved in transecology, actively bringing to the foreground both "trans-" and "Nature" through a close and "transecological" reading of a popular fictional text especially suited to these themes: David Ebershoff's bestselling novel *The Danish Girl* (2000), with some reference to Tom Hooper's Oscar-winning film adaptation of the same name. It should be noted that this chapter will focus predominantly on the novel and is indebted to Ebershoff, not only for the text but also for kindly agreeing to be interviewed for this collection. Ebershoff's novel, which centers on trans woman Lili Elbe, interweaves her story with multilayered images and experiences of the natural world. As such, it lends itself to transecological interpretation, as ideas of "trans-" and the "nonhuman" are intermeshed and brought squarely into the foreground. The text fits in with emerging notions of Queer Ecology, since it serves indeed as a queer reframing of human–environmental relations. Themes of "naturalness," "trans-corporeality," and the very question of what constitutes the "human/nonhuman" resound throughout.

***

The Danish Girl was inspired by real people and real events. It is based on the historical couple Einar and Gerda Wegener (portrayed respectively by Eddie Redmayne and Alicia Vikander in the film version), two successful Danish painters in the early twentieth century. Einar went on to become one of the world's most famous trans women—Lili Ilse Elvenes, more commonly known as Lili Elbe—recognized as a "trans pioneer" (Ebershoff 2000, 315) not only for the fact that she was one of the first individuals to undergo sex reassignment surgery but also for her semi-fictionalized account of her life, which she co-wrote with Niels Hoyer. Published posthumously in 1933, *Man into Woman* is regarded as the earliest autobiographical account of trans- experience and is now a classic in transgender literature (Stone 1992, 224). Although Ebershoff told me that he conducted extensive research into the real-life counterparts of his characters and drew considerable inspiration from *Man into Woman*, *The Danish Girl* is, by his own admission, a largely invented account of the predominantly imagined lives of these individuals. He keeps many of the historical facts the same, but also makes some significant alterations: for instance, Gerda, who in reality was a Danish woman who identified as lesbian, becomes Greta, reimagined as a heterosexual American woman devoted to her husband. In addition, some of Ebershoff's more periphery characters, such as Hans, Einar's childhood friend who he meets again in adulthood, are entirely Ebershoff's own creation. Similarly, while largely faithful to the novel, Hooper also makes use of artistic license in the film version, presenting Elbe, inaccurately, as the very first trans woman to have sex reassignment surgery.[4] It is important to note at this stage that this chapter is concerned with the *character* (as opposed to the historical individual) of Einar Wegener/Lili Elbe and specifically with the ways in which the trans- experience is interconnected in *The Danish Girl* with narratives of the nonhuman. It is important to note that the author, director, and actors involved in the creation and adaptation of "Lili" in both the novel and film do not themselves identify as trans- and are all presenting fictionalized, artistic interpretations of the real-life individual. Although Einar and Lili are not of course two distinct individuals, I choose to echo the language used in both the novel and film throughout this chapter, primarily for reasons of clarity. The name Einar, along with the accompanying male pronouns, is used for the protagonist, while "he" identifies and is seen as a man, and the name Lili, along with the accompanying female pronouns, is used for the protagonist when "she" identifies and is seen as a woman. Such language—inspired in part by the title of the historical work *Man into Woman* (1933)—serves to underline and follow the evolution of the transition of the main character both on-page and on-screen.

The plot of *The Danish Girl* is relatively simple. Essentially, we follow the story of Einar's transition into Lili, and its various consequences. The story begins around the time when Lili is ostensibly first "awakened" in Einar and unfolds as she is given increasing expression. We observe as Einar begins to dress as and slowly "become" Lili, and witness her subsequent, increasingly desperate desire to undergo sex reassignment surgery. Although this physical transformation is ultimately achieved, it is clear at the end of the novel that Lili is about to die as a result of it—and in the film we are shown her death.[5] Throughout the linear narrative we are also given access, intermittently, to the protagonist's past. Significantly, we are immersed in childhood memories of the ambient landscapes of a small place in Denmark called Bluetooth in Jutland. Here, as a child, the protagonist first experiences the feelings of both love and sensuality with the young Hans, but these are forcibly suppressed. We discover increasingly that both Einar and Lili somehow derive and decline—they are both "born" (18) and "buried" (112)—in the bogs of this terrain. The novel features several intriguing and ambiguous allusions to "bones in the bog" (207): a transecological equivalent, perhaps, to skeletons in the closet.

The nonhuman in its various forms—from the sprawling landscapes of Jutland to the various flora and fauna of both real and imagined spaces—has a firmly palpable presence throughout the novel. It is bound, most significantly, to Einar's *transition* into Lili. Extending beyond mere projections of pathetic fallacy, the nonhuman in this text seems not only to reflect but also at times to actively inform the protagonist's transforming relationships to and transitions between the constructions of "masculine" and "feminine. There is a sense of related fluidity between masculine and feminine, and the human and nonhuman. This ties in with Seymour's understanding of "trans-" as defined by its non-stasis and this further relates rather interestingly to Ebershoff's own views on transecology. When in interview I introduced him to this term, he responded with enthusiasm, immediately drawing connections between trans- identities and the natural world. He highlighted the constant if incremental movement of evolution, relating this fluidity to his ideas of trans- experiences:

> Evolution especially speaks to transgender themes—the idea that a species must adapt in order to thrive. The idea that there is no future without transformation. A compelling example: the idea that a creature—a cat, a gazelle, a dragonfly—knows innately what it is. No one can tell the cat that he/she is a dragonfly. The cat acts on instinct. Outside pressure, bias, or ignorance cannot undo the cat's understanding of who he/she is. I believe knowing oneself innately is central to the experience of being transgender.

Whether or not one agrees here with Ebershoff's ideas, they of course hold true inside the fictionalized realm of *The Danish Girl*. Here, when she eventually fully emerges, Lili indeed knows "innately" who she is, and just as Ebershoff draws upon nonhuman examples above to illustrate his point, her self-knowledge is deeply grounded, somewhat literally, in the nonhuman world.

The most central nonhuman conceit of the text, as intimated in the title of this chapter, is of course the protagonist's childhood landscape, which is known simply as "the bog" throughout. The bog is primarily seen in visualizations and flashbacks, but nonetheless steadily permeates the narrative. It serves as the most obvious example of the fluidity and interconnection of the human—and moreover, the specifically *trans-* human—with the nonhuman world, as this landscape is intimately intertwined with the protagonist's transitional experience. In line with the deliberate emphasis in Queer Ecology on *openness* (O'Rourke 2008, xx), the bog is described as a vast, unenclosed, seemingly endless landscape. It is a beautiful, freeing space in which "anything could happen" (198), and so it definitely carries obvious potential for positive symbolism. However, the "open, bubbling mud" also threatens as something unknown and dangerous into which one can "fall" and "slip neatly away" (35). The bog predominantly signifies the struggles involved in the protagonist's repression and burying of the "natural" self within. It stands as a fitting image for Einar and Lili's "muddied" origins as well as for the promise and threat of secrets, with their capacity to either free or drown, which lie hidden beneath the surface.

Both Einar and Lili are closely linked with the wetlands of Jutland. There is an increasingly apparent, if complicated, symbiosis between Einar/Lili and the environment, as the two indeed are presented as somehow *inside* of each other: the protagonist, we are told on numerous occasions, is "in" the bog, "born" in it and "buried" inside; and simultaneously, this terrain somehow resides within the protagonist, as exemplified in the line from the film "*the bog is in me*." Although the bog is bound to the past, it also haunts the present. Although geographically left behind, the environment of his childhood follows Einar into adulthood: this "landscape of the mind" (Melbye 2010, 2) manifests psychically as the near-spectral reflection, if not embodiment of Lili, hidden beneath.

It is in these flashbacks to childhood that we encounter the most detailed and significant description of the protagonist's relationship to the landscape, in which this natural setting's substitution for the trans- self is most plainly revealed. As a boy, Einar plays a game of make-believe with his childhood friend Hans, in which they take the parts of "mummy" and "daddy" respectively, with Einar "dressed up" in an apron to denote femininity. The two children kiss, and within moments Einar's father, enraged, discovers them in this act. Hans manages to escape out into the open landscape, but Einar is caught in the house and beaten. Deeply ashamed of the event, Einar later takes the apron worn in the game, along with various crockery from the kitchen in which the scene played out, and throws it into the depths of the bog, in which it "drowns" (291). This scene is hugely important in its conflation of character and environment. The items, comprising women's clothing and kitchenware, signify gendered ideas of feminine domesticity. Furthermore, they more specifically represent Lili herself as well as her first emergence. The fact that they are so personified—they do not sink, but "drown"—emphasizes her forced submergence. The landscape, therefore, thus becomes such a site of secrets: a graveyard of sorts, concealing the "bones in the bog" that wait to be found. Moreover, the fact that the protagonist's biological sex is conveyed in conjunction with images of emergence, submergence, and landscape echoes Ah-King and Hayward's (2013, 1) sentiments that "sex might be better understood as a dynamic emergence with environment, habitat, and ecosystem." There is the intimation, too, that it is not only the protagonist's gender but also sexuality that is repressed in this formative episode: it is the kiss as much as the attire that marks the memory. This idea that the protagonist's sexual desires have been similarly buried is supported much later on in a scene where we see Lili's sexual awakening as an adult female. Here, when she performs fellatio on a stranger before being interrupted (145), her sensual arousal is coupled with her mind being flooded by images of the Danish wetlands. As the nonhuman seems inextricably intermeshed with human desire, we have a potential example of what Michael Morris has called "eco sexuality": an ecocentric "ecological perspective of sexuality" (Anderson et al. 2012, 99).

Einar, as an adult, is obsessed with the landscape of his childhood. This obsession with Nature—and most especially with the bog—is mediated through art. He is a highly successful landscape painter, renowned for his endless paintings of Danish wetlands. Ebershoff explained to me that Einar is "deeply drawn" in his painting to the environment: he does not paint "people, interiors, or cityscapes," only "the natural world." The subject of the quiet Danish countryside—"the open fields, the bogs, the seas"—appeals because out here "the lid to the soul is lifted" and one can "hear one's own heart more clearly." Ebershoff insisted that much of Lili's transition involves her becoming the person "she already is" and "has always been" and that this inevitably requires an interrogation and understanding of "the natural world she comes from." However, to begin with, as Einar, such interrogation is unconscious. While he presents and identifies as a man, he is obsessed with his artistic subject. Interestingly, Hooper's adaptation in some ways makes the logic of Einar's artistic repetition more explicit: in the film, Vikander's character Gerda draws attention to the fact that Redmayne's Einar paints "the same thing over and over again" and even exhibits some jealousy over his obsession, discussing the bog, ironically, in terms of "the other woman." In endlessly painting the bog, Einar is able in some sense to return to it, to explore this space that he knows, on some level, contains secrets and truths to be revealed. In re-creating this landscape in art, he is able to give some expression to Lili, which is why the task so engrosses him. However, because her expression is not consciously known—she is mediated not only through the landscape but also through the representation of the landscape—Einar is destined to repeatedly return unsatisfactorily to his art. Interestingly, in his most celebrated paintings, he depicts several trees emerging from the bog, symbolizing the possibility of Lili's ascension from the depths of this terrain—a connection further emphasized in the film when Lili is later told by Gerda that she is not welcome and Einar subsequently paints the trees as shrunken and barren.

As an adult, Lili is in a sense "born" from art. In an iconic scene—used in both the novel and the film, and said to be lifted from reality—Greta asks Einar to serve as a substitute for one of her models and pose for a painting in women's attire. This situation seems, at first, to be little more than a game. However, in keeping with the scene from childhood with Hans and the apron, the seeming façade of femininity becomes quickly authentic. This scene importantly occurs in both the novel and the film immediately after we learn of Einar and Greta's inability to conceive a child: one birth, therefore, is symbolically substituted with another. Once Lili is "born," she usurps the role of the bog as she now becomes Einar's greatest obsession. His renditions of Nature are quickly displaced as *she* becomes his extraordinary creation. Now she is expressed in reality—and not merely mediated through landscape on an artist's canvas—Einar no longer has the need, or even the desire, to paint, and Lili never paints at all. The idea that Lili is merged with and then *replaces* the Nature in his art is ingeniously literalized in the film: in a series of beautiful shots, we see that Einar's transition into Lili is largely achieved through *painting* her face. It is not Greta's makeup that is used but Einar's palette, as Lili herself becomes the canvas. There is not the same sense here, however, of mediated Nature, since Lili is shown as authentic and wholly "natural" (4) from the outset.

The protagonist's transition, as an adult, from Einar into Lili is profoundly interconnected with images of nonhuman Nature. In addition to her emergence via Nature in art, there are numerous imbrications between her own personal experiences and more obviously eco-centric perspectives. Indeed she, far more than Einar who likely experiences gender dysphoria, the feeling of being trapped in the "wrong body" (Stone 1992, 228), is presented as "at one" with the natural world. Her very name, "Lili Elbe" which she chooses for herself,[6] is wholly inspired by Nature. Deriving from the flower and the river, her names underline her connections to the nonhuman. Unlike Einar, she is, at least at first, comparatively free in her flesh, able to *feel* the nonhuman world within and around her (she touches and senses the natural, feeling, for example, the texture of petals and the sun on her skin (4)). In the scene in which Einar poses for the painting and Lili is tangibly brought to the surface, her sensed presence, which marks the start of her adult transition, is immediately coupled with images from the natural world. He sees, in his mind's eye, the seemingly incongruous image of "a fox chasing a field mouse" (8, 12) and simultaneously imagines that he hears "the soft cry of a scared little girl" (12). The animals are in conflict and represent different elements of the protagonist's own nature warring within. Meanwhile, the cry of the child surely belongs to Lili, who at this stage has only ever been seen as a *girl* as opposed to a woman. In line with the interconnected symbiosis which exists between Lili and Nature, it is ambiguous as to whether it is her presence that summons the images of the animals, or their visualized existence which awakens Lili's manifestation.

The intimacy that exists between Lili and nonhuman Nature is further underlined as she slowly enters society. On the first occasion when she ventures beyond Einar and Greta's apartment she attends an artist's ball. Although here she is Lili, she is Einar too: uneasy and frightened in what feels, at this stage, like a precarious disguise. It is on this evening, significantly, that she hears tell of something called "the wishing tree" (54). Lili wanders away from the party and soon finds herself in the presence of an enormous oak tree. She is then told by a stranger: "if you eat its acorns you can make a wish and become anyone you want" (54). "If I were to give you an acorn," the stranger continues, "who would you want to be?" (55). Thrown by both the question and by the gentleman's advances, she answers meekly, "I have no idea" (55). Ebershoff told me that he believes the image of the wishing tree, though he said it is now somewhat vague in his mind, was at the time of writing inspired by a Scandinavian myth that he heard long ago. He was captivated by the image and its possibilities, remarking, "who hasn't wished to become someone else, even for a brief period of time?" It is significant that this image is presented while Einar is still riddled with doubts as to who or what exactly Lili is. His true wish—to become Lili permanently and entirely—is as yet unspoken. The tree, in its majesty, silently represents, and seems even to "hear" this wish. The nonhuman, thus, both reflects and informs what lies within the trans- human. Its acorns denote the overwhelming yet magical potential held inside each and every individual. The image of the tree as a symbol of freedom and authentic identity, moreover, continues the symbolism of the trees in the paintings of Bluetooth, rising out of the bog as Lili evolves, continuing to bind the protagonist's trans- self with Nature.

When it becomes clear to others that Lili is to be a central—and tenacious—presence in Einar's world, he is sent to a series of doctors to be "corrected." He is told, again and again, that he is sick, wrong, and unnatural. The abysmal view that to be trans- is to be "monstrous," "outside of nature," "less than fully human," and ultimately a "mistake" (Stryker 1994, 244–5; Elbe 1933) is here firmly and resoundingly echoed. Yet, throughout this period of unwanted medical intervention, Ebershoff draws upon numerous images of nonhuman Nature to emphasize the fact that Lili, perhaps more than anyone around her, is decidedly *natural* in order to undermine these prejudices. He conflates Lili with various living things in order to demonstrate what O'Rourke (2008, viii) calls "interspecies intimacy." When she doubts and suppresses her own identity, she is juxtaposed with visions of harmed and vulnerable Nature, which reflect her own sense of helplessness as well as the "plasticity" in Queer Ecology as human and nonhuman "inform each other" (O'Rourke 2008, xx). For example, when she feigns naïveté about the importance and reality of Lili, the focus moves immediately to a butcher's window, in which small animals, such as piglets, have been skinned (76). She identifies, here, with the weak, unprotected, and violated creatures too quickly consumed by an unthinking and voracious world. Her inner crisis is again seen to manifest in Nature when she is told by doctors that the only explanation for "Lili" is madness. Einar, they decree, is the only reality: he is experiencing schizophrenic psychosis and this is why he appears as "two separate people" (171). In processing this clumsy analysis, the protagonist looks out of the window and sees in the road a dead Alsatian that has been recently hit by a passing vehicle. The dead animal, which is described in direct juxtaposition with Einar's question "do you really believe I'm insane?" (172), is beautifully emblematic of the doctors' thoroughly destabilizing explanation for Lili. In viewing her as mere fantasy, the psychiatrists assert that Lili has never in fact lived, and in this moment, therefore, she is symbolically killed. Consequently, the dog, which has been so bloodily and so suddenly killed in a collision, is Nature's suitable reflection, or perhaps premonition, of the sudden impact upon Einar of this diagnosis. The fact that this view is accompanied by the violent (and anthropogenic) death of one of Nature's creatures serves to suggest that it is in fact the psychiatrists, and *not* the so-called "transsexual," who sin against Nature.

Soon enough, the doctors deem it necessary to physically intervene in order to "cure" Einar. This involves "treatment" via the use of a rudimentary X-ray, which is intended to destroy the "bad" (read: feminine) and retain the "good" (read: masculine) elements within him. The scene is a crucial interrogation, indeed, of "naturalness." Although the procedure is designed to eliminate that which "defies" Nature, it is the X-ray—along with the "science" behind it—that is presented as monstrously unnatural. The "damned machine" is a huge and deafening manmade structure, which "whirs," "clatters," "whips," and "roars" (107–9). In stark contrast, the depths of the protagonist are associated only with the natural, as we are presented with an image of poetic trans-corporeality. We see from Einar's perspective as he imagines what the X-ray will reveal. Echoing Einar's line in the film, "the bog is in me," he sees elements of the landscapes of his childhood buried within. He sees "a stomach alive with glow worms nested from the Bluetooth bog" (108). The imagery here is significant. First, in line with Queer Ecology's call to show the human as "material and relational" (Anderson et al. 2012, 100), the protagonist's body is shown indeed an "an assemblage of materialization in a vast continuum of life and matter." Second, these insects are of course associated with metamorphosis: one of Nature's most obvious and evocative examples of transition, they situate Einar's transition into Lili firmly "*inside* Nature" and echo Seymour's (2017, 257) sentiments of "organic transgenderism" as akin to Nature's different "life cycles."[7] Third, there is something triumphant in the vision of these explicitly *living* and bioluminescent creatures. And finally, the use of the term "nested" as well as the general focus on the abdominal region, which is roughly where the womb lies in pregnancy, further emphasizes not only the fact that Lili is innately female her identity, as Stryker (2006, 10) describes trans- identity, is "ontologically inescapable"—but the promise that she will not remain perennially out of sight, beneath the surface, but will emerge fully following her symbolic gestation.

Eventually, Lili escapes the clutches of unwanted medical intervention and seeks her own doctors, on her own terms. She moves to Paris with Greta, where her anonymity allows her to live a comparatively free and unchallenged existence. Although she no longer presents as or answers to Einar, she grows increasingly desperate and depressed as her physical sex is a daily reminder of what she deems to be her own inauthenticity. Repulsed by her anatomy, she vows to end her existence if it cannot be changed. Yearning for the biology of a woman, she wishes increasingly for motherhood, which she deems—in line with many ecofeminists (as discussed and complicated in Stearney 2009)—the ultimate example of natural femininity, and thus is devastated by the harsh and inescapable fact that "no-one can make a man pregnant" (286). Because of this, she comes to view herself as unnatural and "outside of Nature" indeed. She describes Einar's existence, along with his remaining physical form, as "nature's gravest mishap" (281), echoing the words in *Man into Woman*, where the real Lili Elbe described her own situation as "nature's serious mistake" (Elbe 1933).

In reaction to Lili's increased self-loathing, those who love her search desperately for a solution, until finally they find a man named Dr. Bolk. Bolk assures them that with a series of operations he will be able to alter Lili's biological sex and even create the potential for her to have children. She grows obsessed with this promise and is certain that if the operations are successful she will be able to prove that she is truly a woman: Nature's "mistake," she insists, will be "corrected once and for all" (281). Unlike her previous doctors, Bolk, who is on side with her transition, is associated not only with Nature but also with her relationship to Nature specifically; when he is first introduced he is likened to a "buried animal" (213), echoing her affinity with the bogs and with the significance of what lies hidden beneath the surface. Dr. Bolk promises that Lili will undergo a series of operations that will transform her, physically and entirely, into a biological woman. She is told the final operation will take place in spring: much like the natural world around her, she will surface and be reborn in this season. Although her first series of operations are successful, the final operation of the uterus transplant, as was the case with the real Lili Elbe, proves ultimately fatal. It is performed too early and so Lili is too weak to endure it. We see ominous portents in Nature of her ensuing death. When she is in the women's hospital awaiting the operation, there is a scene in the novel where she is in the garden and we are given a description of its beautiful yet delicate crocuses. These are the only flowers at this stage in the garden and are described simply, but ominously, as "early" (224). By this stage, Lili has been associated on several occasions with flowers (in addition to her name, she is likened to the delicacy of petals (182)) and here once again she is interconnected with the flora. The following day Lili notices them again, as well as the fact that they have multiplied and "spread like a rash" (225). The reader is led, subconsciously, to thus associate prematurity with infection. The earliness of Lili's operation therefore, along with her subsequent death, is illustratively foreshadowed in the natural world.

The endings to the novel and the film are somewhat different. In each, however, the natural environment is significantly intertwined with Lili's fate. In the novel we do not actually see Lili die, but her death is imminent. Although she is in a weakened state following her final operation, she insists "it is too beautiful not to be outside" (306), and so her friends take her beyond the grounds of the hospital to the nearest park. We are given vivid descriptions of the Nature that surrounds her: the grass, the willows, the elms, and the sunlight (306–10). Lili asks if she can be taken "down to the Elbe" (307), the river after which she is named. Now she is, by her own standards, entirely a woman, she can fully embrace her own nature, and this is somewhat literalized as she seeks out her namesake. She is no longer associated with the stagnant muddiness of the bogs of Denmark, which she now "only vaguely" (290) recalls, but instead with the clear, flowing, and pure water of the river. Nor is she bound, any longer, to images of submergence; the final image, which starkly contrasts with any notions of being "buried in a bog," is of a soaring, transcendent kite high in the sky.

In the film, the overall sentiment of the ending remains the same, but Lili's spiritual intimacy with the natural world is perhaps even more explicitly underlined. Here, she dies on-screen, but this does not provide the novel's close. Instead, in the final shots we see Gerda and Hans return to Bluetooth following her death in order to symbolically bid her goodbye. Here, the bog remains an essential element, right to the end, of who Lili is. Greta and Hans return specifically to the part of the bog that in the film Einar was shown to have painted the most: the part in which five lone trees are emerging from the muddy terrain. These trees, as I have suggested above, symbolize Lili's emergence from her origins, from Einar, and from the bog. In the film, we do not have the scene in which little Einar "drowns" the metonymic apron—therein burying "Lili" in the bog—but we do have a symbolic equivalent. In place of the apron we have instead a silken scarf, which is passed continually between Gerda and Lili. The scarf comes to symbolize the feminine in Einar, which is nurtured by Gerda, and so comes to symbolize Lili herself. It is important, therefore, that Greta is again wearing this scarf in the final scene and significant that it is then taken from her by the wind. The very final image is of the scarf, flying high above the trees, much like the kite in the novel, as Gerda speaks the final words: "let her fly free." The "lid of her soul" has been lifted indeed. The idea, therefore, of Lili's emergence from the bog, and all that it signifies, is here even more triumphant. The image of her rising majestically—now "entirely herself"—is powerfully augmented by the addition of mountains to this scene. There are no mountains in the novel (nor, indeed, in Denmark) and the inclusion of this incongruous landscape is therefore a conscious and considered choice by the director to transmogrify the environment to artistic effect. The film ends, therefore, with an unforgettable picture—a near-painting, in itself—of Lili and the landscape. Here, liberated from the confines of her body, she truly becomes "at one" with Nature. In death, she has transcended her physical form, becoming, in line with O'Rourke's (2008, xviii) terms of "queering the non/human," "singular plural": "not substantial, settled, or stable" but one among many "beings in a relational regime independent of identitarianism or anthropomorphism."

Through an examination of the ways in which the human protagonist's transition from Einar to Lili is intermeshed with the nonhuman—and explicitly *foregrounding* the themes centred on "trans-" and "Nature" in the text—we discover a transecological approach to and interpretation of *The Danish Girl*. The novel weaves together the experiences and metamorphoses of the human and nonhuman, resulting in an interconnection of both anthro- and eco-centric perspectives—providing a "queer reframing" indeed of "human–environmental relations" (Brown 2012, 95). This story of Lili Elbe, which as a bestselling novel and star-studded blockbuster movie has reached millions, importantly engages with questions of "naturalness" and "constructedness," and asks us to consider the connections between what it might mean to be "trans-" and what it might mean to be "human." Indeed, the *relationality* (Stryker, Currah, and Moore 2008) of "trans-"—as always moving, crossing, and interconnecting—can teach us a lot about our concepts of the "human" and "nonhuman." Just as Lili is not "outside of nature" (Stryker 1994, 244–5), but within it, with it within her, so too are we inextricably and trans-corporeally intermeshed with the world around us. We, too, echoing Ebershoff's words in an interview about Lili's journey, must interrogate and understand the natural world we come from—not as mere setting but as a part of who we are, that both informs and is informed by us. We must move toward a "new perspective" (O'Rourke 2008, viii), seeing with "fresh eyes" (Hogan, 2012, 87) the world around us. We must thoroughly rethink our relationship to the nonhuman, realizing indeed that we are "multiple" (Luciano and Chen 2015, 186) and "more-than-human" (Seymour, 2017, 255): understanding, as Ebershoff warns, that "a species must adapt in order to thrive" as "there is no future without transformation."

Notes

1 The question of what exactly constitutes "Nature"—and whether humans should be considered inside or outside of it—is complex to say the least. The intricacies of the debates on this subject are not the focus of this chapter but are interestingly discussed by Gail Grossman Freyne (Chapter 9, this volume) under the heading "What is Nature?" In this chapter I use the term "Nature" in broad terms, meaning the collective and nonhuman natural world.

2 I use "trans-" as opposed to "trans" throughout this chapter in line with Susan Stryker, Paisley Currah, and Lisa Jean Moore (2008), who explain in a special issue of *Women's Studies Quarterly* that the "-" is used to emphasize the *relationality* of trans-.

3 Julia Kuznetski (Chapter 4, this volume) focuses in detail on trans-corporeality.

4 In fact, the first was Dora Richter in 1922, and it was the success of her orchidectomy and vaginoplasty that encouraged Elbe with her own operations.

5 Both the novel and the film reflect the fact that the real Lili Elbe initially had successful sex-reassignment surgeries but died from complications following her uterus transplant operation.

6 This is the case in Ebershoff's novel and Hooper's film, but in reality the name "Elbe" was given to Lili by a journalist. In the film, the symbolic potential of the surname is underlined when Lili is first asked for her surname, and after a moment she says, "Elbe … *like the river*."

7 There is, of course, a wider debate concerning the "naturalness" of being transgender and the "reinscribing" of trans- identities as natural is not an uncontested issue. In this volume, both Gail Grossman Freyne (Chapter 9) and Wibke Straube (Chapter 3) explore these specific ideas in more detail.

References

Ah-King, Malin and Eva Hayward. 2013. "Toxic Sexes: Perverting Pollution and Queering Hormone Disruption." In *O-Zone: A Journal of Objectoriented Studies* 1: 1–12.

Alaimo, Stacy. 2010. *Bodily Natures: Science, Environment, and the Material Self.* Indiana: Indiana University Press.

Anderson, Jill E., Robert Azzarello, Gavin Brown, Katie Hogan, Gordon Brent Ingram, Michael J. Morris, and Joshua Stephens. 2012. "Queer Ecology: A Roundtable Discussion". Hosted by Jamie Heckert. *European Journal of Ecopsychology* 3: 82–103.

Azzarello, Robert. 2008. "Unnatural Predators: Queer Theory Meets Environmental Studies in Bram Stoker's *Dracula*." In *Queering the Non/Human*, edited by Noreen Giffney and Myra J. Hird, 137–57. Burlington, VT: Ashgate.

Butler, Judith. 1990, 2007. *Gender Trouble*. New York: Routledge.

Ebershoff, David. 2000. *The Danish Girl*. London: Weidenfeld and Nicolson.

Elbe, Lili. 1933. *Man into Woman: An Authentic Record of a Change of Sex*. London: Beacon Library.

Gaard, Greta. 2010. "Green, Pink, and Lavender: Banishing Ecophobia Through Queer Ecologies." Review of *Queer Ecologies: Sex, Nature, Politics, Desire*. *Ethics & the Environment* 16(2): 115–26.

Giffney, Noreen and Myra Hird. 2008. "Introduction. In *Queering the Non/Human*, edited by Noreen Giffney and Myra J. Hird, 1–16. Burlington, VT: Ashgate.

Luciano, Dana and Mel Y. Chen. 2015. "Has the Queer Ever Been Human?" *GLQ: A Journal of Lesbian and Gay Studies* 21(2–3): 184–207.

Melbye, David. 2010. *Landscape Allegory in Cinema: From Wilderness to Wasteland.* New York: Palgrave Macmillan.

O'Rourke, Michael. 2008. "Series Editor's Preface: The Open." In *Queering the Non/Human*, edited by Noreen Giffney and Myra J. Hird, xvii–xxi. Burlington, VT: Ashgate.

Roberts, Suzanne L. 2008. "The EcoGothic: Pastoral Ideologies in the Gendered Gothic Landscape." PhD diss., University of Nevada.

Scharff, Virginia J. 2003. *Seeing Nature through Gender*. Lawrence: University of Kansas Press.

Seymour, Nicole. 2014. "Trans Ecology and the Transgender Road Narrative." In *The Oxford Handbook of Ecocriticism*, edited by Greg Garrard. Oxford: Oxford University Press.

Seymour, Nicole. 2017. "Transgender Environments." In *The Routledge Handbook of Gender and Environment*, edited by Sherilyn MacGregor, 253–69. London: Routledge.

Stearney, Lynn M. 2009. "Femininsm, Ecofeminism, and the Maternal Archetype: Motherhood as a Feminine Universal." *Communication Quarterly* 42(2): 145–59.

Stone, Sandy. 1992. "The Empire Strikes Back: A Posttranssexual Manifesto." *Camera Obscura* 10: 150–76.

Stryker, Susan. 1994. "My Words to Victor Frankenstein Above the Village of Chamounix: Performing Transgender Rage." *GLQ* 1(4): 237–54.

Stryker, Susan. 2006. "(De)Subjugated Knowledges: An Introduction to Transgender Studies." In *The Transgender Studies Reader*, edited by Susan Stryker and Stephen Whittle, 1–18. London: Routledge.
Stryker, Susan, Paisley Currah, and Lisa Jean Moore. 2008. "Trans-, Trans, or Transgender?" Special issue, "Trans." *Women Studies Quarterly* 36(3–4): 11–22.

Filmography

Hooper, Tom. *The Danish Girl*. 2016. USA: Universal Pictures.

2 Coming out, camping out

Transparent's eco-ethical approach to gender

Katherine Thorsteinson and Hee-Jung Serenity Joo

"Let's go into the forest and menstruate on a stick!" Fusing the images of femininity, biology, and nature, these fem-punk lyrics beset trans woman Maura Pfefferman (played by Jeffrey Tambor) as she sets up camp with her two daughters at Idyllwild Wimmin's Music Festival. The wilderness offers these radical lesbian feminists an escape from patriarchal streets, offices, and homes. Yet, according to the festival's "womyn-born-womyn" policy, Maura is simply the wrong *kind* of woman. This exclusion from the L of LGBT parallels a similar misidentification with the 'T' in a 1994 flashback scene when Maura attends Camp Camellia, a forested "sleep-away camp for crossdressers." She soon discovers the camp's strict policy against hormones and its pervasive culture of misogyny. Glasses are raised when one camper declares, "We are crossdressers, but we are still men!" Within this campy subculture, nature offers a wild space for freedom and subversion, but Maura is simply the wrong *kind* of trans. So go the episodes "Man on the Land" (2015) and "Best New Girl" (2014) from Amazon Studios' award-winning dark comedy, *Transparent*.[1] As the show recurrently emphasizes, Maura's position as a trans woman exceeds both biology and performance. She is thus excluded from Romanticism's vacillating deployments of nature as a measure of purity, authenticity, and gender norms, but also wild(er)ness, freedom, and the radically supra-cultural. Significantly, Maura's identity is most hopefully asserted and hurtfully rejected when she camps in these nature spaces.

As a whole, the series negotiates the "complexities of oppression and privilege" that have contrived to "fracture feminist and queer communities along identity fault lines" (Heyes 2013, 202). While many critics have praised *Transparent* for navigating these inter- and intra-group tensions with careful honesty (if not always successfully), most have overlooked how the series puts these questions into a distinctly ecological perspective. As we will outline throughout this chapter, Timothy Morton's (2007, 2010) critique of "Nature"[2] and preference for a language of "ecology" is reflected in the show's rejection of naturalizing discourse and distrust of designated Nature spaces. Moreover, his concept of the "mesh" (Morton 2010), which loosely translates to ecological entanglement, accurately captures how these episodes approach the ethics of identity claims——insisting on the uncertainty, relationality, and volatility of gender categories. Because our existence necessarily entails our *co*-existence in the infinitely vast and diverse mesh of all things, Morton argues for an ethics of "intimacy" over "inclusion." Translated into a trans discourse, this model can avoid the limits of inclusion politics that ultimately do not challenge hierarchies of power. In what follows, we build on these initial observations to formulate what we consider a transecological theory about the intersecting concepts of gender and Nature.

Throughout this chapter, we thus challenge the politics of "inclusion" structuring radical, liberal, and eco feminisms alike. This goal prompts us to make several historical and methodological diversions. After briefly outlining the gendered histories of North American wilderness camping and the anti-trans fractures that have occurred within feminist Nature spaces, we review some recent trans theory and ecocriticism in order to explore the potentials for cross-fertilization. Finally, in our close readings of "Man on the Land" and "Best New Girl," we latch on to the various homonyms of "camp" as a framework for exploring the tensions between Nature and gender. For example, flashbacks to the Pfeffermans' European Jewish ancestry recall the Nazi concentration camps that put questions of biological purity into brutal perspective. In the same episode, questions of racial privilege and settler colonialism are subtly raised when Maura shops for owl feather earrings at the marketplace and interacts with a woman named Vicki who is perhaps Native American. So, too, when Maura's femininity is mistakenly read as drag or kink, the show reveals how ironic camp aesthetics can do just as much to reinscribe as to subvert gender essentialism. For Morton (2007, 98), however, irony is the most ecological register because it "involves distancing and displacement, a moving from place to place, or even from homey place into lonely space." Indeed, irony is not merely a comedic device but an attempt to grasp things as they really are—aporetic, multiple, transforming, what Morton (2013) calls "realist magic."

By exploring these and other resonances, camp emerges as an assemblage of contradictions: natural and constructed, spatial and doctrinal, transient and entrenched, communal and militant. Citing Eva Hayward and Jami Weinstein in her Preface to this volume, Greta Gaard (p. xx) similarly reminds us of how " 'trans* is not a thing or being, it is rather the processes through which thingness and beingness are constituted … trans* troubles ontologized states.' " We thus take our cluster framework of "camp" as a potential trans methodology for reading laterally, intimately, and (im)provis(at)ionally. We wonder: What new meanings can be produced by traversing these connotative linkages? Can this homonymic argumentation provide a model for an "eco-ethics" of intimacy over inclusion, sensitivity over sameness? And might this strategy of horizontal rather than hierarchical argumentation attend to some of the troubling fault lines that have emerged between and within queer, feminist, and trans politics?

Indeed, we take our non-linear and highly contingent methodology to be a practice in transecological thought. But asking these questions also prompts us to "come out" about our own subject positions and motivations for writing this chapter. As feminist and queer scholars who unequivocally reject trans-exclusive radical "feminism," we must also respond to the tensions that are increasingly pressing us to think about what "us" in fact means. We do not intend to speak on behalf of trans people in general or trans women in particular and have avoided conjectures about others' experiences altogether. Our thoughts are very much indebted to several transfeminist theorists and activists—we defer our readers to them for questions that exceed our scope and capacity.[3] While our focus here is on how trans women are interpellated within feminist and queer frameworks, we hope that our transecological approach will also be relevant for thinking trans men and masculinity or questions of group difference more broadly.[4]

~~~

In their introduction to *Queer Ecologies*, Catriona Mortimer-Sandilands and Bruce Erickson (2010) explore the modern history of wilderness camping in the United States to exemplify how particular Nature spaces and sexualities came to be mutually reinforcing. In the late nineteenth century, urban cores became sites of increasing economic independence for women with the reorganization of patriarchal family relations, the restructuring of employment under capitalism, and the changes in ethno-racial demographics. To escape the emasculating effects of industrialization and immigration, urban bourgeois men retreated into the wilderness to recuperate their sense of power. By extension, many thought "the dominant social body [could be restored] through rigorous, health-giving recreation" (Mortimer-Sandilands and Erickson 2010, 13). Then, following America's economic boom after World War II, camping increasingly became a car-based leisure activity. The desire to push women back into heterosexual domesticity following their involvement with the war effort also helped promote this auto-recreation culture, now less focused on masculine virility than on middle-class family values.[5] Complicit in the naturalization of white male supremacy, heteronormativity, and cis-patriarchy, the early parks movement had pronounced biopolitical overtones.[6]

By the 1970s, however, as feminist activists and scholars were beginning to theorize about the interconnections between different structural oppressions and networks of power, many attempted to reclaim Nature from the grasp of masculinity and heteronormativity. The back-to-the-land movements of this era articulated new environmentalist ideals that were intimately bound up with what were then considered to be radical queer and feminist politics. Lesbian separatists recognized the patriarchal organization of society as responsible for most of America's problems, and they turned toward Nature as an alternative to male-dominated cities and suburbia (Unger 2010). For example, the southern Oregon Womanshare Collective was founded on the idea that "women could find, 'in the healing beauty of nature,' 'a safe space to live, to work, [and] to help create the women's culture [they] dreamed of'" (Unger 2010, 28). While many of these Nature communities were built for permanent dwelling, others were formed as more provisional retreats and music festivals. These seasonal safe havens enabled women to make professional and political advancements in
~~~

urban centers, yet still return annually to "safe" and "nurturing" spaces (Unger 2010, 186–7).

According to Mortimer-Sandilands and Erickson (2010, 28), whatever their temporal or social organization,

> These "wimmin's lands" had complex ecological goals, ranging from opening rural landscapes to women by transforming heterosexual relations of property ownership, to withdrawing the land from patriarchal-capitalist agricultural production and reproduction, to symbolically reinscribing the land with lesbian erotic presence, to creating a distinct lesbian "public sphere" founded on both lesbian separatist and overtly ecological concerns.

On the one hand, the creation of queer rural spaces resisted the late nineteenth-century view of Nature as inherently heteromasculine and also reclaimed the natural world from middle-class domesticity. As early as the nineteenth century, in fact, lesbian authors were using "pastoral literary traditions to develop a reverse discourse that argued for the naturalness of women's same-sex love relationships and/or the congenital equality of lesbians" (Mortimer-Sandilands and Erickson 2010, 24). These back-to-the-land movements thus largely extend from this longer history of reverse Nature discourse.

On the other hand, some of the political problems that prompted this feminist turn to Nature bear similarities to those of the late nineteenth century—in particular, fears of urbanization. Built upon settler colonialist assumptions of "empty" land, back-to-the-land movements were also often rooted in heterosexual assumptions regarding Nature as a place untainted by urban civilization. Replacing heteromasculinity's rugged wilderness were ideas of Mother Nature and Mother Earth that celebrated essentialist notions of female fertility, reproduction, and care. At the same time that these movements were claimed as feminist safe spaces of solace and celebration, they either ignored Native American displacement or romanticized Native cultures (or both). Nancy Unger (2010, 173, emphasis added) explains that lesbians created these alternative environments in an effort to "transcend the sexism, homophobia, violence, materialism, and environmental abuse *afflicting* mainstream society." This language of "affliction" echoes in Sandilands' description as well:

> [R]ural separatists viewed the land as a place that could restore physical and spiritual health to a group of people sickened, literally, by (heteropatriarchal capitalist) corruption and pollution and thus as a sort of paradise on earth to which women could be admitted if they recognized their oppression at the hands, and in the hands, of men.
>
> (quoted in Unger 2010, 181)

This pathologizing rhetoric certainly diagnoses a different social ailment than late nineteenth-century fears of emasculation, but it nonetheless assumes certain troubling dichotomies—Nature/culture, feminine/masculine, health/disease—that have reproduced other kinds of erasures and exclusions, in particular for trans people.

In this back-to-the-land feminist view, cities figure as the centers of toxic masculinity. Rape and violent crime, corporate glass ceilings and sexual harassment, capitalist competition, and unrealistic body ideals indicative of hyperconsumerism all capture this sense of urban hostility toward women. Yet within this schema, patriarchy has been too easily conflated with maleness writ large and the symbolic phallus has been reduced to the functioning penis. Accordingly, the city is aligned with maleness and masculinity, while Nature corresponds with femaleness and femininity. The problems with this type of thinking are revealed most clearly by trans-exclusionary radical feminists (TERFs) who simplistically and incorrectly view trans women as male and thus as enemy oppressors. As seen in *Transparent*, this type of spatial reordering persists into the twenty-first century and continues to leave trans women literally with no place to be.

For example, "Man on the Land" is likely inspired by true events that have occurred at the (in)famous Michigan Womyn's Music Festival. Held every August since 1976, the festival folded in 2015 amid criticism against its trans-exclusionary policies. In theory, organizers prided themselves on "welcoming women of all nationalities, ages, races, sexualities, and physical abilities" (Unger 2010, 187). Despite insisting that diversity was "strongly valued," however, Unger (2010, 187, 189) admits—rather perfunctorily—that the festival was only "open to all women-born-women." She clarifies (parenthetically) what this term means: "those who were born and raised as girls and who identify as women, excluding transsexual and transgender women—one of several policies generating heated debate within the queer community" (Unger 2010, 189).[7] The policy was "instated in the early 1990s after an incident in which a woman named Nancy Burkholder was expelled from the festival" when it was suspected that she was trans (Serano 2007, 234). The gender determinism of this policy was also reflected in some of the festival's events, such as the woman-centered spiritual practices that emphasized "women's 'oneness' with the earth, with the moon, and with natural cycles" (Unger 2010, 188). The biological essentialism of this rhetoric is flagrant: women's bodies, menstruation, and "life-giving-forces" were thought to align them more closely with Nature.[8]

We claim it is no coincidence that such battles over "real" or "natural" womanhood occurred in Nature spaces. Rather than serving as a mere backdrop, the very idea of Nature—including assumptions about the "natural"—forced separatist feminists to face their own limitations.[9] In fact, the very disputes about transness that arose in these Nature spaces prompted new ecological relationships. For example, following her expulsion from the festival, Nancy Burkholder and three other trans women returned in 1993 with allies to "conduct scheduled Festival workshops and outreach to attendees" but were turned away once again by security (Koyama 2003, 7–8). They retreated to the national park across from the main gate where they camped in the woods for several days. Throughout this period, they conducted their workshops with 75 women and met with 200 others. Thus, through what Heather Love (2007, 20) calls "the reversibility of reverse discourse," Camp Trans was born. While the back-to-the-land movement reversed heteropatriarchal claims to Nature, Camp Trans reversed the gender exclusions inherent to even lesbian feminist discourse.

Surviving several years of hiatus and some organizational changes, Camp Trans protested until 2010 when the Michigan Festival officially withdrew their policy and incorporated groups such as "Trans Womyn Belong Here" and other trans-focused workshops, even though the space continued to be occupied by many TERFs (Ring 2015). Camp Trans thus became an important space in trans political organizing. However, beyond this successful mobilization of wilderness space, trans women have been excluded for contradictory reasons in much Nature discourse. They are either regarded as not really women, demonstrated by *Transparent*'s Idyllwild Wimmin's Music Festival, or they are assumed to be too committed to their identities as real women, seen in Camp Camellia. The reversibility of reverse discourse reveals itself to be a slippery tactic, equally appropriable for cross-purposes. This ambivalence arises because reverse Nature discourse—whether feminist, queer, or trans—always relies on the dichotomy between inclusion and exclusion, even as it redefines the specific relation between these terms. A transecological framework asks instead that we rethink the false dichotomies structuring these reversals, aiming for a politics of intimacy over one of inclusion.

Julia Serano (2013, 117) argues along these lines that theories of gender have been limited by the false dichotomy between Nature and culture or, in her words, between "determinism" and "artifactualism." The conflations of gender determinism with "biology" and gender artifactualism with "social constructionism," she argues, have been responsible for the doubled and contradictory exclusions of trans people.[10] According to Serano, this false dichotomy between the biological and the social explains why trans women are almost invariably construed as unnatural in both gender determinist and artifactualist accounts. Determinists expect that sex/gender should conform to their simplified understandings of Nature, whereas artifactualists view the sex/gender binary as itself wholly fictional or "unnatural."

Talia Mae Bettcher argues similarly that trans narratives have been unnecessarily limited by two conflicting approaches to gender identity: the "wrong-body theory" and the "beyond-the-binary model." On the one hand, the wrong-body theory assumes a "misalignment between gender identity and the sexed body" (Bettcher 2012, 383).[11] Whether this misalignment is considered a condition of the mind or a problem of the body, this theory encourages what Bettcher calls "reality enforcement" whereby "real" sex and/or gender is thought to be empirically verifiable.[12] Trans people are thus imagined to be either "pretenders" or "deceivers," pathologized for their non-normative gender identities and marginalized for their "wrong" bodies. The wrong-body theory thus unfairly places the burdens of exclusion and oppression on trans people themselves instead of critiquing the social structures that produce these gender hierarchies in the first place.

On the other hand, the "beyond-the-binary model" aims to subvert the dichotomy between the categories "male" and "female" by unifying all non-normative genders (including transsexuals, nonbinary and gender fluid folks, cross-dressers, drag queens and kings, and others) under the umbrella term "transgender."[13] However, the view that gender is a mere cultural construction, which seemingly underpins this model, misrepresents the actual experiences of many trans people for whom "gender identity seems impervious to cultural modification" (Prosser 1998, 5; Bettcher 2012, 385). Many trans people identify forever and entirely with one gender, and the beyond-the-binary model invalidates these identities by suggesting they are naively mistaken. From this perspective, the "self"-identifications of trans persons themselves are politically problematic because they reaffirm the oppressive gender binary (Namaste 2005, 7; Bettcher 2012, 385).

Feeling caught between the Scylla of the beyond-the-binary model and the Charybdis of the wrong-body framework, Bettcher (2012, 404) concludes that we are "trapped in the wrong theory" and calls for a version of constructionism that is more attuned to the "nonreducibility of trans oppression." While we should continue to theorize gender as constructed and contingent, we must also be attuned to the performative registers of trans resistance and, indeed, the significance of personal experience or identification amid allegations of inauthenticity. This also requires that we recognize "multiple worlds of sense" in which particular gender practices may have different consequences, and terms such as "woman" may resonate with multiple meanings (Bettcher 2012, 403). For example, while liberating in some queer and feminist circles, the argument that everybody's sex and gender are constructed ignores and thus quietly reinforces the specific ways in which trans people are "constructed *as constructions*" (Bettcher 2012, 398, emphasis in original).

Dismissing trans claims to innate gender identity from a cis perspective is thus a categorical mistake. Trans people who clearly identify as men or women deploy these identity categories in ways that radically depart from dominant patriarchal *and* second-wave feminist gender practices, and by doing so they alter the very meanings of these terms. At the same time, their claims to gender essentialism are not as incompatible with constructionism as some feminists have come to believe. Indeed, we must simply turn to theories of performativity to distinguish "between actual acts of resistance"—which might affirm gender realness—and "a theory that illuminates that resistance"—which may reveal how *all* deployments of gender are constructed (Bettcher 2012, 398). This distinction reinforces Anna Bedford's (Introduction, p. 7, this volume) reading of Judith Butler: "gender can be performative without being a mere performance." We all must ask, though, how Bettcher's "multiple worlds of sense" resonate with the growing ecological awareness that we are all in a very literal sense part of the *same* world.

We answer this question by way of the term "transecology." We might think of gender, that is, less as a binary or even as a spectrum and more as a rich ecology—an ecology which is now reconfiguring in spectacular ways with the pressures and potentials afforded by the growth in trans discourse, as well as other recent social and political changes. Transecology is thus a gender politics that does not assume it already knows what gender is (and isn't). Instead of reiterating a gendered essentialism onto Nature (as either ruggedly masculine or nurturing and feminine), a focus on how we interact with such ideas about Nature can help us rethink human hierarchies, including gender divisions, altogether. *Because* we all share the same world, our deployments and inhabitations of gender will necessarily shape each other in incredibly intimate and often inharmonious ways. Trans-exclusionary radical feminists attempt to "settle" (with all of the colonial and agrilogistic resonances of that word) the rich and raucous gender ecology, literally and figuratively defending their female "turf" from perceived outsiders. The mainstream liberal focus on "inclusion" thus capitulates to the proprietary and individualistic TERF framework by soothing fears about gender sovereignty.

The inside/outside structure of "inclusion" will not facilitate trans justice, just as it cannot save the Earth from manmade catastrophes. As Julia Serano (2013, 200) warns, "when we single out some force outside ourselves … [i.e. a predefined notion of 'man'] as the source of the marginalization we face, it encourages us-versus-them thinking." As ecocritic Timothy Morton (2010, 274, emphasis added) similarly explains, "ideologies of *Nature* are founded on inside-outside structures that resemble the boundaries heterosexism policies." Traditional environmentalists, for example, often encourage humans to put aside their differences in the face of the world's vastness and bond together to save a Nature that exists "out there," outside of us. Seen from an *ecological* perspective, though, all "life-forms, along with the environments they compose and inhabit, defy boundaries of inside and outside at every level" (Morton 2010, 274). Contemplating this shift from Nature to ecology requires a shift in scale, from anthropocentric reordering to reconceptualizing the limits and boundaries of the category "human" itself in relation, not only to each other, but also to the rest of the nonhuman world. Likewise, the existence of trans women—an existence better described as tenacious perseverance within this violently transphobic world—must be welcomed as an opportunity, not to pathologize "trans" but to problematize "woman." Since such pathologizing discourses have often pivoted around the human/nonhuman binary, a deliberate turn to ecology as a larger system that does not prioritize humans can provide alternative ways to think about gender justice. Given these points of convergence between ecological and trans theories, then, we can begin to trace an ethics and language for identity claims that will usefully inform and cohere transfeminist politics.

This project begins with Morton's (2012, 78) claim in *The Ecological Thought* that we need "a vision of intimacy" rather than "a vision of inclusion." To apprehend this vision, Morton (2012, 29–30) develops his metaphor of "the mesh" to capture the "infinite connections and infinitesimal differences" between all living and non-living things: "All life forms are the mesh, and so are all the dead ones, as are their habitats, which are also made up of living and nonliving beings." Everything is interrelated negatively and differentially in a totalizing open system without central position or ontological hierarchy (Morton 2012, 40, 38). Morton (2012, 28) explains that "mesh" is preferable to other terms such as "network" or "assemblage" because it means both "the holes" and the "threading between them." Indeed, focusing on the interdependence of all beings complicates the boundaries between interior and exterior, foreground and background, existent and nonexistent (Morton 2012, 39, 28). In other words, this framework spans difference but "permits no distance," entailing a radical and paradoxical intimacy that differs markedly from liberal rhetorics of inclusion (Morton 2012, 39).

Intimacy in the mesh makes the familiar strange and the strange familiar, both dissolving and expanding our sense of self. Indeed, the more we know about something, the more ambiguous it becomes. Morton calls these ambiguously inscribed beings "strange strangers." We can never know the strange stranger until we meet them. Even then, "they are liable to change before our eyes, and our view of them is also labile" (Morton 2012, 40). They are beyond the scope of language or comprehension, and they take many forms—"After all, they might be *us*" (Morton 2012, 41, emphasis added).[14] Thus, unlike "interconnection" which "implies separateness and difference" (Morton 2012, 47), this intimacy dissolves the distinctions between self and other such that two important things happen. We come "to terms with the passivity and void of the strange stranger" at the very same time that we realize "the strange stranger is us, [so] the void is us, too" (Morton 2012, 80). For Morton (2012, 41, 80), intimacy with strange strangers is thus "threatening" and "disturbing," yet our very "basis for reimagining democracy." This is what we call "eco-ethics": a form of care that does not strictly distinguish self from other and—unlike liberal rhetorics of inclusion—"is about coexistence but not about harmony" (Morton 2012, 66).

Our concept of eco-ethics reverberates with the etymological kinships that Susan Stryker (Foreword, p. xviii, this volume) notes: "ethics," "ethos," "oikos," and "eco." She concludes from this tangled genealogy that "Ethics ... is inherently transecological, and must be attuned to histories—and emergences—of embodied difference in all its forms, including those (perhaps especially those) that problematize the nature/culture duality." Translated to a trans framework, then, this eco-ethical intimacy with the strange stranger could be seen as a twofold illumination: First, debating whether or not trans women are women is the very problem. Second, the real problem lies in ideas of "woman" in general. It is not that "trans" is strange to "woman," which would only reinforce transphobic notions of pathology and abnormality. It is rather that "woman" has always already been strange to herself. Indeed, Morton (2012, 80, 42, emphasis added) tells us that "the strange stranger *is us*" and is also "something or someone whose existence we cannot anticipate." Likewise, trans studies critic J. Halberstam (1994, 226) emphasizes "the strangeness of all gendered bodies." Thus, although " 'transgender' is a word that has come into widespread use only in the past couple decades [and] its meanings are still under construction" (Stryker 2008, 1), its histories and potentials are inherent to the very concept of gender itself.

In their contributions to this volume, Susan Stryker (Foreword), Anna Bedford (Introduction), and Peter I-min Huang (Chapter 5) demonstrate how concepts of gender are "haunted" by "transgender phenomena" across diverse cultures. And Gail Grossman Freyne (Chapter 9) suggests that the socially constructed binaries of gender *as well as* sex entail that we are all simultaneously "masculine and feminine," "male and female"—even while cis and trans people embody this simultaneity differently. Eco-ethics thus calls us to uncanny anachronisms: Imagine all those women lost to history, written in the books as men. Imagine all those women lost to themselves, having written themselves out as men for lack of a different language. Now that we have this language for trans identity/experience/phenomena (a language that has always existed in some sense and in some spaces), we can begin to trace ourselves through these histories, potentials, and one-time impossibles. This is what it means to be intimate with strange strangers, to find ourselves (anew).[15]

~~~

This practice in intimacy can also lead to defensive territorialism. Indeed, Morton describes this intimacy as "uncanny," even sometimes "threatening" or "disturbing." These negative affects dominate the tonal landscape of *Transparent*. A long-time divorcee, retiree, and "Moppa" to three grown children, Maura has come out to her family and friends later in life. In part due to her age, she finds herself almost literally "transparent" when she moves to a lively gay apartment complex that seems to favor youth, masculinity, and loud music. This sense of isolation brought on by the gay community is also reinforced by her own family. In one heartbreaking scene, Maura is abandoned by her children on the "Trans Got Talent" stage while performing Goyte's apposite song "Somebody That I Used to Know" (a title that may initially seem to reference Maura's gender transition but later appears to describe her family's altered affections). Yet neither is Maura entirely blameless, especially when it comes to her own class prejudices.

She receives an earful after insulting her friend Davina's (played by Alexandra Billings) patronizing and possessive boytoy. "We don't all have your family, we don't all have your money," Davina candidly replies, "I'm a fifty-three-year-old ex-prostitute HIV-positive woman with a dick. And I know what I want, and I know what I need." Finally, there are some profoundly complicated scenes revealing Maura's own ageism—a sexual preference for much younger women—and her ready if unconscious acceptance of male privilege in the past—excluding women from a Berkeley editorial board as an undergraduate student and taking the family house following divorce thanks to the convention of signing deeds in the man's name. As Sonia Saraiya (2015, n.p.) writes, *Transparent* "is not didactic, but it does not pander, either … [in a single moment] it is both about the most intimate dealings between humans and also the grand ideas that move them." Like Morton's ecological thought, the show finds an ethics of intimacy in our vast mesh of social relations and resists the easy one-size-fits-all rhetoric of "inclusion" offered by mainstream LGBT politics.
~~~

Indeed, the show launches a heavy critique of liberal capitalist self-actualization and individual-rights discourse. For example, after Maura's daughter Sarah recklessly breaks off a series of romantic relationships, her stilted fiancée Tammy (played by Melora Hardin) drunkenly crashes the Pfeffermans' pool party where she finds her wedding cake half eaten by guests. "You think there are no consequences?" Tammy pleads, "I'm a fucking consequence! I'm not crazy! I'm in pain! *I am your pain.*" Dissolving the inside-outside distinction that structures egoism,[16] Tammy reveals how "autonomous" choices have consequences that necessarily affect others. Indeed, so entangled are their actions that these "consequences" are better articulated as identifications: "*I am your pain.*" Both Bettcher and Morton critique the consumerist language of "choice" motivating and justifying Sarah's actions. For Bettcher, "choice" inaccurately describes the process toward gender affirmation as mere *whim* and, for Morton, it masks the highly eco-ethical nature of human action as simply *personal right* (rather than "enmeshed" relation). Maura's other children also frequently make rash and selfish decisions—Ali loses her childhood friend and lover Syd when she equivocates over and then ultimately resists monogamy, while Josh loses his fiancée when he fails to mourn the miscarriage of their baby. Although this egoism devastates the lives of nearly every secondary character, an eco-ethics of the mesh can be observed in the relationships among the Pfeffermans themselves. Hardly the picture-perfect nuclear family, they somehow bare and bear their vulnerabilities together until the end. Certainly, it is not biology that keeps the family together, despite the fact that it is what binds them.

The creative process, aesthetic form, and narrative content of *Transparent* may also help distinguish the mesh from models of inclusion such as "assemblage," "mosaic," or "network." Jill Soloway has called their[17] directorial style the "female gaze": it "is about creating the conditions for inspiration to flourish, and then 'discerning-receiving'" (quoted in Saraiya 2015, n.p.). Critic Melissa Silverstein speculates that the female gaze is not "simply the reverse of the male gaze … [it] is not about pleasure or even power; it is about *presence*" (quoted in Saraiya 2015, n.p., emphasis in original). Soloway suggests that this presence is neither decisive nor forceful but is rather—like Jewishness and femininity—"centered around questions" (quoted in Saraiya 2015, n.p.). "[W]e live in a world dominated by perspectives," they state, "that are overwhelmingly answer-oriented" (Soloway quoted in Kamen 2014, n.p.). Questions have an absorptive presence, existing in themselves and also drawing forth the enigmatic possibility of answers. Morton similarly describes the mesh as feminine because, like questions do of answers, it encompasses both the "threading" and the "holes." Soloway's female gaze is thus also deeply trans, in that "feminine" not only names a way of being but also *asks for* new becomings.

This concept of the mesh is also evident in certain camera techniques and casting decisions that serve to dissolve clear distinction between characters. Near the end of the pilot episode, Sarah brings Tammy to her childhood home under the pretense of asking about interior design. The two have not seen each other since they dated in college and have both since married. Our anticipation increases as they flirt and slowly move toward each other, but at the very moment when they might have kissed the screen suddenly switches to Josh having sex with his girlfriend. The graphic match cut trades in superficial details—blonde hair for blonde hair, brown hair for brown hair—as well as playing with our affective and narrative expectations. Through the lens of LGBT inclusion, homo desire easily translates into hetero sex ("love is love," you'll barely notice the difference). A more transecological reading of intimacy would see how a cisgender sister *becomes* her cisgender brother whereby "transgender" emerges at the intersection of family resemblance, (sexual) desire, visual editing, and cultural expectation. Soloway employs these techniques frequently, creating a stitched-together aesthetic that brings disparate things into relation.

The soundtrack also functions to link our affective response to different episodes, repeating certain songs across various scenes that bear no obvious connection. To similar effect, actors often play multiple roles from different moments in the family's history such that Gaby Hoffmann plays Ali and her grandmother Rose, Emily Robinson plays a younger Ali and a younger Rose, and so forth. The family's genealogical entanglement thus becomes literal; the present is not merely *caused by* or *similar to* the past but is instead *woven together* with the past. The recurring use of mirrors as scene props also contributes to this aesthetic of repetition and relationality. Jay Prosser (1998, 100) has observed that "mirror scenes … constitute a convention of transsexual autobiography. They recur across the texts in strikingly similar fashion." This mirror imagery draws us into a logic of identification *through* opposition (our mirror repetitions are also our opposites). Just as the female gaze does not simply reverse the male gaze, trans women challenge the assumption that "feminine" is merely the derivative and inverse of "masculine." These trans aesthetics thus ask us to refigure not only the gender binary but also the entire logic of reversal structuring both exclusion and inclusion (in particular, lesbian feminist reverse Nature discourses).[18]

Beyond cinematic process and aesthetic, *Transparent*'s circular narrative confounds expectations for linear plot and character development in the *Bildungsroman* genre conventional of "coming-out." Indeed, as Saraiya (2015, n.p., emphasis in original) puts it, this is a show in which "nothing exactly *happens*. There are some shifts and resettlings, but … much of the story of 'Transparent' is of a family engaged in the slow process of becoming whatever they already were. It ends somewhat as it begins—a point in the middle of a process—and favors the excavation of moments to the mapping of arcs." Saraiya (2015, n.p.) notably describes the narrative as "enveloping" and "weav[ing] into the fabric of the place"—images that approximate Morton's concept of the mesh. She further explains this in ecological (if not perhaps also colonial and agrologistic) terms: "Characters revisit old territory and break new ground, but ultimately exist in about the same plot of land—the Pfefferman homestead, as it were. And in that space, certain moments become indelible" (Saraiya 2015, n.p.). Indeed, the narrative of *Transparent* is much more lateral and spatial than it is linear or teleological. In creative process, aesthetic form, and narrative content, then, the show expresses the intimacy of the mesh.[19] Yet, as Morton has warned, this intimacy is far from harmonious, easy, or inclusionary.

Cressida Heyes (2013) elucidates a framework for transfeminist politics that embraces these productive tensions. Although she does not address Nature or the environment specifically, her approach to transfeminism is highly ecological. In particular, her understanding of gender as "a web of relations in ongoing tension and negotiation" as opposed to a purely individual matter conforms to Morton's concept of the mesh (Heyes 2013, 203). As she explains, the "expression of one gender may limit the possible meanings or opportunities available to others" (Heyes 2013, 203). Rather than advocate a laissez-faire "speak your own truth" type of liberalism, then, Heyes (2013, 203) calls for "an ethics of self-fashioning" whereby we "recognize the discursive limits on individual self-transformation without denying agency to gendered subjects." In other words, our freedom actually depends on our mutual imbrication—our "choices" produce the identities and possibilities of others (Heyes 2013, 203). This eco-ethics is vital to Heyes' project of recuperating a transfeminist theory of gender. For the variety of subject positionings under patriarchal discipline gives way to numerous "strategies of divide and conquer," and she insists this has been most apparent in feminist debates about trans identity (Heyes 2013, 211).[20]

Heyes (2013, 208) also follows Serano, Morton, and Bettcher in insisting that our experiences are no less "real or deeply felt on an individual level" just because we are the "contingent product of large historical dynamics." No one can say that any woman is simply the "hapless product of social shifting" or that she simply "upped and chose to be a lesbian—or a transsexual" (Heyes 2013, 208). Rather, we need to navigate the "complex intermediate space" in which we have all been "thrown into particular subject-positions" yet in which we must also ground our strategies of resistance (Heyes 2013, 208). Thus, like Bettcher, Heyes advocates for a distinction between theory and practice whereby we acknowledge multiple worlds of sense, yet her "net of relations" offers an important caveat. That is, although particular deployments of gender involve different motivations and consequences, they are not self-contained or politically neutral. While this "net of relations" sets certain constraints on individual agency, it also forms the basis for a new transfeminist politics.[21]

For one thing, transness is not the only node of difference within feminism, nor perhaps is it even the most significant. Race, class, sexuality, relation to colonialism, and global region are among the most obvious examples of intragroup tension. If "woman" can usefully capture *some* commonality across these distinctions and if "feminism" can politically unify so many cacophonous voices, then why not also for trans women? For another, there are certain political goals that are shared by practically all feminists such as "weakening the grip of oppressive sex and gender dimorphisms" and the "concomitant devaluing of the lesser terms 'female' and 'feminine'" (Heyes 2013, 202). Finally, possibilities for feminists of all stripes can *grow with* the changing definitions of "woman" that trans discourses ignite. As new possibilities for gender freedom open up to individuals, a whole new field of meaning will be generated over time within which "some identities may eventually cease to exist while others are being created" (Heyes 2013, 202). Most notably, transness rewrites the tired narrative of male exceptionalism and female victimization. Trans women have after all forsaken their supposed "male privilege" and have often faced overwhelming socio-economic obstacles and bodily violence for embracing their femininity. Likewise, trans women challenge the assumption that "feminine" is merely the derivative and inverse of "masculine."

~~~

The ironic register of *Transparent* is perhaps what aligns the series most closely with this eco-ethical approach. According to Morton (2007, 100), irony is the only way to be intimate with strange strangers without "killing them"; that is, "turning them into yourself or into an inanimate object." We might replace "environmentalism" with feminism and "strangers" with (trans) women[22] in the following statement and retain Morton's (2007, 100) basic meaning: "If irony and movement are not part of environmentalism, strangers are in danger of disappearing, exclusion, ostracism, or worse." As previously mentioned, *Transparent*'s ironic vision is not merely for comedic effect but also reorients the viewer into this eco-ethical perspective by forcing us to rethink the *grounds* of our assumptions. The entire episode of "Man on the Land" engages in this affective and aesthetic register: Idyllwild is depicted as both parody and pastiche, resplendent and repulsive. On the one hand, blurring the distinction between television and reality, the episode is something of a historical record and homage. A number of popular feminist performers make cameos, including Indigo Girls, Sia, Peaches, Alice Boman, and queer porn star Jiz Lee (some of whom have publicly supported trans presence at such festivals). Moreover, when the Pfeffermans first arrive, the forest appears to be a utopia of love and energy. Women of every race, age, and shape stream through the camp in various degrees of nudity and hairiness. Many are laughing or dancing, some are lounging in the sun, and others are making out with each other against the trees.

In addition to this nostalgic and utopian vision, however, the show pokes fun at some of the festival's political naiveté and self-aggrandizement, including the commodification of racial and ethnic diversity that reveals its participation—and literal location—in settler colonialism. A pamphlet of events suggests "arriving early" for a tampon-making workshop, presumably because of its enormous popularity. The Pfeffermans struggle to enjoy and discern the ingredients of a vegan-friendly "nutloaf." Later, at "Shaman Crying Bear's Intention Circle," an ostensibly white woman with a thick New York accent and wearing deerskin and headdress stands in a tepee and reminds everyone about her "Drumming Away Racism" group. The foolishness of this scene calls attention to the colonialist appropriation of both Native land and culture as well as the show's own investments in whiteness, exposing the ways in which second-wave feminism has been complicit in settler colonialist logics. After all, whether utopian or satirical, Idyllwild exists as a space for women because it has been deliberately "emptied."
~~~

Maura begins to detect a rift between the appearance of feminist celebration and the reality of feminist exclusion while perusing "ethnic" jewelry in the marketplace. She strikes up a pleasant conversation with Vicki (played by Anjelica Houston), another festival-goer, who informs Maura that the earrings are made from owl feathers. The sounds of a wooden flute playing in the distance increase during their exchange. Vicki has dark, braided hair and dark eyes, wears a cowboy hat and numerous pieces of turquoise jewelry, and is a cheesemonger in Marchmont. Within the shared space of a feminist marketplace crammed with appropriated ethnic goods, viewers are compelled to scrutinize Vicki's "authentic" racial and ethnic identity, parallel to yet not conflated with the ways in which Maura's "authentic" gender identity is under scrutiny. Vicki suspects that Maura is a trans woman and informs her of the "womyn-born-womyn" policy. Although Vicki seems to disavow this "bullshit policy" (her phrasing), Maura flees into what now looks like an angry and prying mob to find her daughters. Significantly, little alters in the appearance and actions of the festival-goers themselves. Instead, the camera moves and cuts more abruptly between images, and the soundtrack emphasizes the disharmonies of drum circles, folk guitar, laughter, and shouting. In other words, it is merely our *perspective* that has shifted. Maura belatedly realizes that she had always been excluded from those previous scenes of utopian revelry.

Lost in the chaos of normalized bodies and potential persecution, Maura attempts to find some quiet and relief at the port-a-potties. But as several workmen clean the facilities, she is disturbed by the eruption of an aggressive chorus repeating "Man on the land! Man on the land!" A rather weather-worn clown named Sherlock explains the function of this chanting: "when the men come, and they only come to take our shit away, we like to say 'man on the land'… so that no one gets triggered or too excited." Sherlock represents everything that Maura has discovered in these trans-exclusionary radical feminists: her wide welcoming smile is rigidly painted, her whitewashed makeup is beginning to fade, and her gender-bending tie and bowler hat are mere costume. As a spokesperson for the festival's policy, Sherlock is doubly ridiculous—utterly unaware that her straight-faced explanation clashes with her slapstick appearance. The name "Sherlock" should, moreover, remind us of the famed detective who is at once a queer eccentric and an unwavering Truth hound for "reality enforcement."

Yet, if the ironic intention of this image seems obvious, the ultimate meanings of several other scenes are much more ambiguous. For example, when Maura finally finds her daughter Ali sitting around a bonfire with other campers she is persuaded to sit down for a beer. Prior to her arrival, a poet in the group reflects upon their shared insignificance: "We are goddamned specks of dust … tonight it feels so damned good just being a nothingness speck with my chosen family." This statement should perk up our ears to the possible eco-ethical intonations of the subsequent encounter. However, after a relatively friendly beginning, the conversation degenerates into what Julia Serano (2013, 200) has called an "oppression Olympics." Several of the women defend the festival's policy of excluding trans women, jumping back and forth between the injustice of male privilege, the triggering effects of penises, and the pressures to discontinue nudity for fear of rape. Maura is at a disadvantage. Hardly can she repeat her question—"Is that what this is about?"—before she must absorb another blow. Indeed, this entrenchment of gender normativity and hierarchy contrasts sharply with the preceding eco-ethical desire for a family of nothingness specks.

The reasoning against Maura is dizzyingly disjointed and takes on various degrees of plausibility. We are likely repulsed by one woman's masculine colonialist rhetoric that she "drove the plow" and "cleared the woods," thus bearing no responsibility for Maura's discomfort. We must also be skeptical of another's claim that every camper had been raped, and of her further insinuation that Maura would engage in rape herself. Yet the encounter gets more complicated when Ali's girlfriend Leslie (played by Cherry Jones) redirects the conversation toward questions of male privilege. Maura protests, "I was in way too much pain to experience what you're calling privilege." "Your pain and your privilege are separate," Leslie tells her. The show continually grapples with this debate over relative privilege and, in line with its ironic aesthetic, no solutions are produced around the campfire. Instead, Maura abruptly leaves the festival, shouting in pain and anger: "This woMAN is leaving this feminist fuckhole! Thank you for your kindness and FUCK YOU!" (Throughout the series, however, viewers are frequently presented with other examples of class and race privilege which do not seem to undermine claims to womanhood.)

Without warning, the narrative lines begin to blur between the Pfeffermans and their ancestors in Nazi Germany. Ali chases after Maura but finds herself running in traditional Ashkenazi shoes, which were described once before in a previous episode. We are then transported back to the Magnus Hirschfeld's Institute for Sexual Research in 1933, where Maura's mother Rose and "Auncle" Gittel (who is also a trans woman) lived in their youth. The Nazis have gathered outside and they begin to raid the building, throwing books into a fire and arresting several gender-variant people. The parallels between the festival and the Institute are overt but complex. On the one hand, this book-burning recalls the bonfire that spurred Maura's escape. The scene oscillates between Gittel's fear at the Institute and Maura's anger as she gathers her belongings. So too, the episode's first extra-diegetic music, Alice Boman's "Waiting," plays in harmony with the Nazi brass band and a festival fiddler, suggesting a rhythmic connection between the gender oppression across these historical moments. Yet, there is no straightforward equivalence between these two narratives either. The brief flashes of intimacy in the succeeding montage complicate the simple literalization of the "feminazi" slur: Ali and Leslie tenderly undress each other in their tent, Vicki picks Maura up in her car and drives her safely away, and Sarah actualizes the BDSM fantasies which seem to supply a new perspective on her gendered self.

Beyond the fact that both episodes explore the connections between gender naturalization and Nature spaces, "Man on the Land" and "Best New Girl" also share several structural similarities. In particular, both use narrative flashbacks and an ironic tone. However, in "Best New Girl," irony takes on a more campy aesthetic that feels misaligned with Maura's sincerity. Until part-way through the episode, we may not be quite aware that Camp Camellia is a retreat for cross-dressers. It is 1994, and we hope with Maura that this will be the moment when she finds her community. Of course, this hope is impossible from the very start, since we understand this trip to have occurred in the past and, as we know, she is yet to feel full "inclusion" in any space. When Maura arrives at the welcome party with her friend Mark (played by Bradley Whitford), who goes by the name of Marcy at camp, our creeping sense of unease with the extravagant clothing and dramatic performances is both tempered and intensified by the confident jocularity of the crowd. Something seems off about the mannerisms, and about the voice inflections. Yet everyone is so kind and free, anticipating the false utopia at Idyllwild Wimmin's Music Festival that trips up Maura again later in her life.

Does this representation transphobically exaggerate gestures and clothing, the viewer may wonder, or does this very suspicion amount to transphobic gender policing? We tighten into a critical posture, but do not know who or what should be the subject of our critique. Then perhaps it hits us, and Maura too, when another party-goer spits, "Girls can be such mean little twats!" Camp Camellia is, of course, a retreat for cross-dressers and many of the campers feel entitled to their (trans) misogynistic views. Later in the episode, Mark and Maura's cycle to the public payphone is a telling scene in this respect. Outfitted in modest summer dresses, the two pedal idyllically through the sun. When he speaks to his unwitting family, however, everything about Mark's demeanor changes: he slouches his back, cracks his neck, clears his throat, deepens his voice, and advises his son: "man-up" to the coach. After witnessing the break in Mark's fourth wall, Maura appears disappointed or shaken and refuses to call home herself. Utopia has been revealed as a mere temporary escape.

Our reactions to this scene may echo Serano's concerns over ironic and campy aesthetics. She is uncomfortable with camp in general because she believes it derides femme gender expression. Echoing Mat Fournier's (Chapter 6, this volume) point that non-female is always assumed to be sexless or neutral, Serano explains that "feminine expression is always viewed as an act, as a performance" (Serano 2013, 63). Yet "Man on the Land" and "Best New Girl" both engage in complex forms of irony that hold onto the multiplicity of these meanings rather than collapse them into mocking admonition. This irony follows from Soloway's question-oriented female gaze; instead of resolving inter- and intra-group tensions with simple answers like "inclusion," *Transparent* opens us up to intimacy in the mesh.

~~~

If we are to think of "inclusion" as answer-oriented and "intimacy" as question-oriented, we may still have a feeling of lingering uncertainty: How *do* we resolve our fraught encampments? Throughout this chapter, we have argued that
~~~

the inside–outside distinctions structuring our ideas of Nature and gender keep hierarchies of power in place. We thus take our eco-ethical approach to provide some self-fulfilling questions (or rhetorical answers), bringing together the often disparate and bellicose fields of ecological, feminist, and trans studies. This transecological framework is also evident in these two episodes, "Man on the Land" and "Best New Girl," as well as in the creative process, aesthetic form, and narrative content of *Transparent* more broadly. Like Morton, the show eschews concepts of the natural or Nature, didacticism, teleology, and individualism, preferring to highlight the uncertainty, relationality, and volatility of gender categories. In other words, instead of trying to "fix" the parts that are not working within a world we have already defined, it may be more useful to understand that the world is continuously being redefined in a perpetual state of flux.

This resolves our problem with a paradox. To be truly intimate with others we must see how their very strangeness is itself strange, a form of defamiliarization that brings us into a closer relationship with the world. Our radical intimacy with strange strangers in the mesh thus provides "a platform for compassion rather than condescending pity" as we begin to see our fundamental entanglement with others (Morton 2012, 80). This intimacy is neither comforting nor peaceful; it is fraught with confusion and disagreement. But such terms have always described the landscape of feminist politics. Indeed, from a transecological perspective, gender is a web of relations in ongoing tension and negotiation. Deployments of gender are never isolated, but there is no objective schema for determining who should conform to which gender ideal or indeed why conformity should be valued in the first place. We must therefore begin with our intimacy: trans women are women. Next, we can begin to ask how our gendered entanglements are shaped and how we can create a world of greater trans*flourishing—a world in which all inhabitations of gender continue to be transformed.[23]

Notes

1 *Transparent* won a Golden Globe for Best Television Series in 2014, becoming the first streaming media series to do so. Tambor received both a Golden Globe and Emmy Award for his performance as Maura, furthering criticism of the persistent casting of non-trans actors in trans roles. Since the initial writing of this article, Tambor was fired in 2018 over allegations of sexual harassment on set.

2 In *The Ecological Thought*, Morton (2012, 3) cautions that he will "sometimes use a capital N to highlight [Nature's] 'unnatural' qualities, namely (but not limited to), hierarchy, authority, harmony, purity, neutrality, and mystery." In much of this chapter, we follow this capitalization to defamiliarize these associations.

3 With special thanks to Joshua Bastian Cole from Cornell's Department of Performing and Media Arts for his attentive feedback on this chapter.

4 Morton (2010, 274) raises a few warnings in his approach to queer ecology that are also relevant here:

> Unfortunately, a great deal of ecocriticism provides a toxic environment in which to spawn queer ecology. Ecofeminism … arose out of feminist separatism,

> wedded to a biological essentialism that, strategic or not, is grounded on binary difference and thus unhelpful for the kinds of difference multiplication that is queer theory's brilliance. Much American ecocriticism is a vector for various masculinity memes, including rugged individualism, a phallic authoritarian sublime, and an allergy to femininity in all its forms.

We are sensitive to these tensions and thus turn to Morton's engagement with the term "queer ecology" as a foundation for developing our notion of transecology.

5 Camping sites were established to resemble suburban culs-de-sac and were separated from "public" activities like hiking, swimming, and climbing. The cultivation of trees ensured this "privacy," though sites usually opened up onto a shared path so that park rangers could "make sure nothing illegal or immoral was taking place" (Mortimer-Sandilands and Erickson 2010, 19).

6 Indeed, it is not coincidental that public parks emerged contemporaneously with notions of evolutionary degeneracy and environmental contamination linked to urban spaces. In fact, many medical thinkers believed that homosexuality was an illness caused by polluted environments (many still do).

7 Yet, while this policy was enforced in several occasions, polling data from festival-goers reveals the deeply held confusions about what "womyn-born-womyn" could actually mean (Koyama 2003, 18). In certain cases, trans men were recognized as men and thus accused by feminists of "joining the enemy." In other instances, trans women could be rejected for their "unnatural" claims to womanhood, or else accepted as women but still rejected on the grounds that they once received male privilege. Even on the level of biology, the policy was both tautological and inaccurate. Released 25 years after Nancy Burkholder's rejection from the Michigan Womyn's Music Festival, *Transparent* provides a satisfactory response back to these logical fallacies. As Ali asks in *Transparent* after learning that the Festival's policy requires a vagina and a uterus, "Does that mean if you've had a hysterectomy then you're not a woman?" Moreover, as studies have shown, "many transsexual women actually feel that they have always been women, albeit misidentifi[ed] by others, and thus feel that the phrase 'womyn-born-womyn' should include them" (Koyama 2003, 5). Yet others under the trans umbrella, such as butch women, bearded women, and drag-king performers, were not excluded by this policy so long as they lived their entire lives as "women" and did not identify as genderqueer. In fact, some festival performers labeled themselves "transgender" under the broader umbrella definition, while at the same time supporting the Festival's exclusionary policy against trans women (Koyama 2003, 5–6). In other cases, "'trannie boys, boydykes, FTM's, Lesbian Avengers and young gender-variant women'—who were not transsexual women—were evicted from the festival for their refusal to identify as women, or in solidarity with other trans people [*sic*]" (Koyama 2003, 6). Such contradictions and inconsistencies further show how fraught the category of woman—or womyn—continues to be. Instead of viewing this chapter of feminist history only as a moment of exclusion from existing definitions of woman, we recognize it as a moment of ontological challenge to these definitions.

8 Moreover, like the auto-recreation culture of the 1950s that sought to foster middle-class family values, these women's music festivals were surprisingly reliant on public–private divisions. Many of those in attendance sought to escape "the material objectification of women in violent U.S. media" and yet embraced nudity in the privacy that forests provided (Unger 2010, 90). In combination with the festival's biologically essentialist theories of gender and trans exclusionary policies, this culture of nudity did not provide freedom but rather increased risk of exposure for trans women. In other words, this claim to privacy was distinctly a cis privilege, and in this sense their culture of nudity is somewhat analogous to the public pathways that allowed park wardens to enforce postwar standards of decency.

9 Unger (2010, 195) nevertheless reminds us that in addition to challenging "sexism, homophobia, and violence, lesbian communities … made important contributions to environmental history and environmental justice movements." Many of these queer and feminist communities still exist, and some continue to be politically fraught.

10 According to social constructionism, gender is not purely determined by biology "but rather is shaped to some extent by culture—by socialization, gender norms, and the gender-related ideology, language, and labels, that constrain and influence our understandings of matter" (Serano 2013, 118). Although often conflated with social constructionists, gender artifactualists take a harder line, rejecting altogether the role of biology and biological variation in shaping our genders. Serano (2013, 139, emphasis in original) acknowledges the appeal of artifactualism for queer and feminist activists who see it as the only alternative to "*gender determinism*—the belief that women and men are born with predetermined sex-specific behaviors and desires." Yet, these activists misinterpret biology as *necessarily* determinist even though the most salient accounts of sex and gender, including Darwin's theory of evolution, in fact emphasize the importance of sociality. Consequently, Serano (2013, 118) warns that even nuanced and qualified "suggestions that biology may have *some* influence on gendered behaviors or desires will garner accusations of 'essentialism' in gender artifactualist circles."

11 In its strong version one's real sex is determined by gender identity as opposed to genitalia or chromosomes, and in its weak form one's sex is made to align with gender through genital reconstruction. Underpinning the fields of sexology, medicine, and psychiatry, this framework is also embraced by some trans people.

12 The four essential features of reality enforcement are identity invalidation, the appearance—reality contrast, the deceiver—pretender double bind, and genital verification:

> Identity invalidation is the erasure of a trans person's gender identity through an opposing categorization (e.g., a trans person sees herself as a woman, but she is categorized as a man). This invalidation is framed in terms of the appearance–reality contrast (e.g., a trans woman may be represented as "really a man disguised as a woman"). And this contrast is manifested in one of two ways that constitute a double-bind for trans people—namely, passing as nontrans (and hence running the risk of exposure as a deceiver) or else being openly trans (and consequently being relegated to a mere pretender). Genital verification can be a literal exposure (as with Brandon Teena, Gwen Araujo, and Angie Zapata) or else a discursive reveal through euphemistic comments like "was discovered to be anatomically male." These disclosures anchor identity invalidation in the notion of genitalia as a kind of concealed reality.
>
> (Bettcher 2012, 392)

13 This theory emerged in the 1990s from trans scholars and activists Sandy Stone, Leslie Feinberg, and Kate Bornstein, among others, in response to the medical industry's pathologizing use of "wrong body" discourse. This account thus developed at a moment when trans people were beginning to theorize themselves for themselves in the face of internal fragmentation and external gender oppression. It also enabled strategic ideological alliances with queer and feminist movements that had already been developing constructionist theories of sex and gender. Subsequently though, thinkers such as Jay Prosser, Henry Rubin, and Viviane K. Namaste have begun to raise serious worries that this vision does not accurately capture the realities of many transgender people.

14 To elucidate this oscillation between familiarity and strangeness, Morton (2012, 36) reminds us that the "insides of organisms teem with aliens." Our cells contain the original Archæan anaerobic bacteria that created the ecological disaster called oxygen. Surgeons transplant organs and scientists grow tissue in laboratories. Bees and flowers evolve together. Our engines run on crushed dinosaur parts and iron is "mostly a by-product of bacterial metabolism. So is oxygen. Mountains can be made of shells

and fossilized bacteria" (Morton 2012, 36, 39, 29). As these examples illustrate, our radical intimacy both "veils the mesh beneath the illusion of familiarity" but also profoundly disturbs our sense of identity through space and time once we become more ecologically aware. In other words, our intimacy with strange strangers is deeply uncanny—if something seems too familiar then we must not be seeing its enormity and extension, but if something seems too strange then we must not be seeing how we ourselves are dependent upon, implicated with, and even constituted by it.

15 Just as intimacy is not "interconnection" or "inclusion," it is also not "equivalence." Cis and trans experiences of gender will be different, but this is something that Black Feminism has already explained in terms of intersectionality. There are many different, even incompatible, ways of being a woman but all of these must be understood to constitute women's experience, identity, positionality, and history.

16 The opposite of eco-ethics, egoism is a liberal capitalist ethics governed by the principles of *harm* and *autonomy*.

17 Soloway "identifies as a gender non-conforming queer person, who … [uses] gender-neutral pronouns (they/them/their)" (Freeman 2017, n.p.).

18 Moreover, trans aesthetic becomes trans gaze with Soloway's *"transfirmative* action program" prioritizing the employment of trans creators and crew. For example, trans producer/director Rhys Ernst has edited the opening credits for each season, Silas Howard has also directed on the show, and trans writers include Ali Liebegott and Our Lady J.

19 For Morton, the mesh is properly inconceivable because it captures the relationality between all things. There is a risk of reducing what exceeds our imagination to metaphor, but the mesh is highly immanent, material, even mundane. We want to take this narrative disruption in our chapter as an opportunity to remind readers of this tendency to abstract from "the mesh" in the analytic mode.

20 Although she would agree with Heyes' general thesis and aims, Julia Serano provides an important qualification. Some might be compelled to read in Heyes "the assumption that we should all curtail or alter our genders and sexualities in order to better conform with feminist or queer politics" (Serano 2013, 5). In other words, if gender is a web of relations, some might think that individuals should moderate themselves to reduce their impact on others. According to this liberal-capitalist ethics weighing harm against autonomy, people may make "choices" until they begin to impede the choices of others. The effects of this logic—what Anne Koedt calls "the perversion of 'the personal is political'"—are that inclusion comes at the cost of assimilation, and individuals (read: minorities) must conform to the interests of the majority. She argues that these one-size-fit-all approaches are common to both reformist and radical feminists. Reformist feminists often seek "to purge 'less desirable' identities and behaviors from their movements in the name of political expediency" while radical feminists denounce "identities and behaviors that they perceive to be too 'conservative,' 'conforming,' or 'heteronormative'" (Serano 2013, 5). In either case, confirming or resisting gender norms, homogeneity is preferred. Significantly, Serano (2013, 5) points to distinctly ecological reasons for thinking beyond this normative impulse: the fact "that there is naturally occurring variation in sex, gender, and sexuality in human populations." It is important to note, however, that Heyes does not reproduce "the perversion of 'the personal is political'" herself.

21 Similarly, we might recall Morton's image of the strange strangers meeting in the mesh to reimagine democracy.

22 We use the slightly awkward parentheses here to grasp the specificity of "trans" but also the general category of "woman" simultaneously.

23 Avery Tompkins (2014, 26) explains:

> the asterisk (*), or star, is a symbol with multiple meanings and applications that can mark a bullet point in a list, highlight or draw attention to a particular word or

phrase, indicate a footnote, or operate as a wildcard character in computing and telecommunications. In relation to transgender phenomena, the asterisk is used primarily in the latter sense, to open up transgender or trans to a greater range of meanings.

References

Bettcher, Talia Mae. 2012. "Full-frontal Morality: The Naked Truth about Gender." *Hypatia* 27: 319–337.

Freeman, Hadley. 2017. "Transparent's Jill Soloway: 'The Words Male and Female Describe Who We Used to Be'." *The Guardian*, May 21. Accessed October 11, 2017. www.theguardian.com/tv-and-radio/2017/may/21/transparents-jill-soloway-the-words-male-and-female-describe-who-we-used-to-be.

Halberstam, J. 1994. "F2M: The Making of Female Masculinity." In *The Lesbian Post-modern*, edited by L. Doan, 210–28. New York: Columbia University Press.

Heyes, Cressida. 2013. "Feminist Solidarity after Queer Theory: The Case of Transgender." In *The Transgender Studies Reader 2*, edited by Susan Stryker and Aren Z. Aizura, 201–12. New York: Routledge.

Hird, Myra. 2006. "Animal Transex." *Australian Feminist Studies* 49: 35–50.

Kamen, Paula. 2014. "*Transparent*'s Jill Soloway on Inventing the Female Gaze." *Ms. Magazine*, November 6. Accessed August 31, 2016. http://msmagazine.com/blog/2014/11/06/transparents-jill-soloway-on-inventing-the-female-gaze/.

Koyama, Emi. 2003. *A Handbook on Discussing the Michigan Womyn's Music Festival for Trans Activists and Allies*. Accessed August 31, 2016. http://intermargins.net/repression/deviant/transgender/trans_book/mich-

Morton, Timothy. 2010. "Queer Ecology." *PMLA* 125(2): 273–82.

Love, Heather. 2007. *Feeling Backwards: Loss and the Politics of Queer History*. Cambridge, MA: Harvard University Press.

Mortimer-Sandilands, Catriona and Bruce Erickson. 2010. Introduction to *Queer Ecologies: Sex, Nature, Politics, Desire*, edited by Catriona Mortimer-Sandilands and Bruce Erickson, 1–50. Bloomington: Indiana University Press.

Morton, Timothy. 2007. *Ecology Without Nature: Rethinking Environmental Aesthetics*. Cambridge, MA: Harvard University Press.

Morton, Timothy. 2008. "Ecologocentrism: Unworking Animals." *SubStance* 37(3): 73–96.

Morton, Timothy. 2010. "Queer Ecology." *PMLA* 125(2): 273–82.

Morton, Timothy. 2012. *The Ecological Thought*. Cambridge, MA: Harvard University Press.

Morton, Timothy. 2013a. *Hyperobjects: Philosophy and Ecology after the End of the World*. Minneapolis, MN: University of Minnesota Press.

Morton, Timothy. 2013b. *Realist Magic: Objects, Ontology, Causality*. Ann Arbor, MI: Open Humanities Press. Accessed August 31, 2016. www.openhumanitiespress.org/books/titles/realist-magic/.

Namaste, Viviane. 2005. *Sex Change, Social Change: Reflections on Identity, Institutions, and Imperialism*. Toronto: Women's Press.

Prosser, Jay. 1998. *Second Skins: The Body Narratives of Transsexuality*. New York: Columbia University Press.

Ring, Trudy. 2015. "This Year's Michigan Womyn's Music Festival Will Be the Last." *The Advocate*, April 21. Accessed August 31, 2016. www.advocate.com/michfest/2015/04/21/years-michigan-womyns-music-festival-will-be-last.

Saraiya, Sonya. 2015. "This Is What the Female Gaze Looks Like: *Transparent* Returns for a Thrilling Second Season." *Salon*, December 13. Accessed August 31, 2016. www.salon.com/2015/12/13/this_is_what_the_female_gaze_looks_like_transparent_returns_for_a_thrilling_second_season/.

Serano, Julia. 2007. *Whipping Girl: A Transsexual Woman on Sexism and the Scapegoating of Femininity*. Berkeley, CA: Seal Press.

Serano, Julia. 2013. *Excluded: Making Feminist and Queer Movements More Inclusive*. Berkeley, CA: Seal Press.

Stryker, Susan. 2008. *Transgender History*. Berkeley, CA: Seal Press.

Tompkins, Avery. 2014. "Asterisk." *TSQ: Transgender Studies Quarterly* 1(1–2): 26–7.

Transparent. 2014. "Best New Girl." Season 1, episode 8. Directed by Jill Soloway. Written by Bridget Bedard and Jill Soloway. Amazon Prime, September 16.

Transparent. 2015. "Man on the Land." Season 2, episode 9. Directed by Jill Soloway. Written by Jill Soloway. Amazon Prime, December 11.

Unger, Nancy C. 2010. "From Juke Joints to Sisterspace: The Role of Nature in Lesbian Alternative Environments in the United States." In *Queer Ecologies: Sex, Nature, Politics, Desire*, edited by Catriona Mortimer-Sandilands and Bruce Erickson, 173–98. Bloomington: Indiana University Press.

3 Posthuman ecological intimacy, waste, and the trans body in *Nånting måste gå sönder* (2014)

Wibke Straube

A body is always constituted by multiple other bodies: some live inside of us as, for instance, bacteria, while many others are in close proximity, such as other humans, animals, and further organic and inorganic materialities: "We are not human alone—we are human with many" as Eva Hayward cunningly summarizes (2008, 70). In this chapter I will examine the multiple co-becomings of human bodies with their intimate Others. In particular I will discuss the notion of "posthuman ecological intimacy," which will facilitate my analysis of art and film through a transecological lens. This term will function as the backbone of my discussion of the transfeminine main character Ellie in the Swedish film *Nånting måste gå sönder* (*Something Must Break*, Dir.: Ester Martin Bergsmark, 2014), its transecological visual enactment of environmental pollution, and the strong aestheticization of waste in the film.[1]

Intimacy beyond a human-centered understanding is culturally rarely discussed, and yet it presents one of the main tropes of the film. It is an intimacy that entangles the character Ellie with an always present, polluted nature. Much in opposition to a general understanding of contaminated landscapes, the film allows garbage to become an aesthetic entity. In this film waste is not an absolute. Waste itself is generally dependent on its context (Douglas 2002;Hird 2012): a discarded empty plastic bottle can within weeks reappear as a designer T-shirt or a vertical garden project, but it may also end up contributing to the Great Pacific garbage patch. Mostly, things become rendered as waste when their usefulness has expired or they are thrown away, taken out of sight, dumped into landfills, shot into space, or shipped off to far regions. Hence, in *Nånting*, nature is never enacted without pollution. The dichotomy of clean versus polluted is eliminated. In addition, nature is not connected to a pastoral or rural countryside. It appears in various places, surburban, rural as much as urban, sometimes in the form of a deserted playground, or as a remote and overgrown amusement park, Stockholm's harbor, semi-urban wastelands (Figure 3.1), a forest lake next to a highway ramp, and in the form of sprouting weeds in all possible spaces (Figure 3.2).

This is the scenery in which the film unfolds the story of its two main characters Ellie and Andreas. This is where they meet, fall in love, and break up: forgotten landscapes, cruising areas, and remote, and mostly overgrown, semi-industrial

Figure 3.1 Högdalstoppen 1, a hill as semi-urban wasteland in Stockholm.

Source: Still from *Nånting måste gå sönder*, Ester Martin Bergsmark (director), Lisabi Fridell (cinematographer), Ester Martin Bergsmark/Eli Levén (script). Copyright: Garagefilm International AB.

Figure 3.2 Close-up of weeds exemplary of the film's intimate depictions of nature and range of close-ups of waste and discard matter.

Source: Still from *Nånting måste gå sönder*, Ester Martin Bergsmark (director), Ester Martin Bergsmark (cinematographer), Ester Martin Bergsmark/Eli Levén (script). Copyright: Garagefilm International AB.

suburbanity. This non-dualistic approach to nature and the position of the trans body within seems to question any attempt of dualistic boundary demarcations between nature/wilderness and industrialized urbanity, and also how these places are rendered culturally: nature and the "outdoors" as a historically masculine and heterosexually coded space versus the urban, especially metropolitan space as historically linked to sexual, racial, and political diversity as well as normative transgression and "deviancy" (Mortimer-Sandilands 2005). Yet, nature's gendered ecology is ambivalent. It is also in ecofeminist traditions the space that is coded as feminine, as the one equally exposed to toxic masculinity and its violence. In this chapter, I discuss how the film *Nånting* reconceptualizes the gendering and coding of this space through an enactment of transecological, anarchist aesthetics. Partly, it reworks the polarized coding by chosing semi-industrialized, suburban sites which are neither fully the metropolican center nor the rural countryside, and reappropriates nature as a space of queer contaminedness. Furthermore, it engages in a daring aestheticization of pollution and links it to the trans/ing body.

The term "trans" (or also "transing") is conceptualized as a verb in my writing. "Trans" relates to gender as well as extending beyond it. In its gendered meaning, historian Susan Stryker has defined trans as a "movement across a socially imposed boundary away from an unchosen starting place" (Stryker 2008, 1). While "trans" or "transgender" are on the one hand very specific gender identities based on the experience of medically and/or socially transitioning, both are also sometimes used as umbrella terms that include a multiplicity of gender nonconforming positions (e.g., transgender, transsexual, cross-dressing, intersex, nonbinary, agender, butch, and more). I understand it in particular as a verb that indicates a movement away from birth-assigned sex (Enke 2012, 5) as well as a "practice that takes place within, as well as across or between, gendered spaces" (Stryker, Currah, and Moore 2008, 13).To follow this quotation, the term "trans," in my rendering, includes gender as well as opening up toward further transformatory practices in humans and their intimate Others.

Returning to the discussion of the film *Nånting* in this chapter, I will investigate how the human trans body is in this film presented as intimately entangled with organic and inorganic pollution and, literally, garbage. By drawing upon a trans and queer ecofeminist and ecocritical framework I discuss what implications this entanglement has for an understanding of intimacy in posthumanist feminist ontologies as well as for trans bodies in the context of nature and contamination (Cole 2016; Gaard 1997; Hayward 2008; Hird 2012; Mortimer-Sandilands and Erickson 2010; Morton 2010; Oppermann 2013; Woelfle-Erskine 2016, n.d.; see also Thorsteinson and Joo (Chapter 2, this volume)). Accordingly, I will first explore how the film's particular aesthetic strategies incite different forms of intimacy that include more-than-human entities on equal terms with the human trans body. Next, I will inquire into which ways certain aestheticizations of waste in the film subvert and ironically play with a notion of "clean" or "pristine" nature. Finally, I will examine how the enactment of polluted nature in the film initiates unexpected encounters and affinities between different intimate Others.

In a Western context the trans body, similar to the queer body, has a history of being culturally associated with terms such as "unnatural," "impure," or "polluted." For instance, this is evident in the terminology "gender identity disorder" used in *The Diagnostic and Statistical Manual of Mental Disorders* (prior to the most recent edition where this has been revised) which addresses trans embodiment as a mental illness (Winters 2005). The trans body has also been perceived as "less than human" and excluded from the category of the human in its transgression of gender norms and normative sexuality. In addition, the white, working-class body has been positioned outside of the category of the human as "wasteful" and, for instance, visible in the slur, degraded as "white trash." Moreover, dehumanized human bodies often find themselves on the outskirts of the "right" values of capitalist productivity, health, and the expected upward social mobility (Mortimer-Sandilands 2005; Stryker 2006; Chen 2012). In these problematic and materially devastating renderings, the trans body has been understood not only as impure but also as "contaminating" itself (Haraway 1997; Stryker 2006). This is mirrored, for instance, in the ongoing practices of compulsive sterilization of trans bodies in many parts of the world—including until recently supposedly liberal countries such as Sweden, Denmark, and Germany.[2] The trans body is ontologically squeezed into proximity to what in Western contexts is understood as "unnatural." "Unnatural" is here also synonymous with "unhealthy." The powerful discourses of Western medicine and psychiatry have defined the trans body as defying standards of health, and as socially, mentally, and physically "deviant" and "pathological" (Butler 2004; Spade 2006; Burke 2011; Krieg 2013). Consequently, the trans body has historically been stigmatized as "unhealthy," "unnatural," and "polluted." The film *Nånting* and its aesthetics can then be read as unfolding a surprising politics in the face of these stigmatizing discourses. Instead of dismissing the declassification of the trans body as an "unnatural" body, the film unfolds an anarchist approach of appropriation rather than dismissal of the "stigma," which would ultimately result in transnormative assimilation. In opposition, the film's aesthetics embrace the position of pollution, even of toxicity (Engel and Lorenz 2013; Straube 2019), and create an arena of shared intimate Otherness that extends an invitation to all subjects and entities, declassified as "impure" and "polluting."

The backbone of the discussion in this chapter is the concept of intimacy with Others–this is a posthuman ecological intimacy. To briefly introduce the concept, my deployment of the term *intimacy* draws upon Stacy Alaimo's notion of trans-corporeality as a contact zone between human and the more-than-human nature (2008, 238; see also discussions by Parker (Chapter 1, this volume) and Kuznetski (Chapter 4, this volume). The terms *posthuman* and *ecological* draw upon the feminist critiques of anthropocentricism (Åsberg et al. 2011; Oppermann 2013). *Ecological* in particular emphasizes the focus on nature and pollution, environmental justice, and environmental ethics. I will discuss this in more detail in the following pages.

This chapter approaches the discussion of the film *Nånting* through a framework of transecology and transecological aesthetics. In previous publications, I have applied transecology as a reading lens for artistic material, as it allows a conceptual frame for the entanglement of ecology, nature, trans embodiment, and transing bodies more broadly (Straube 2019a, 2019b). Ecocritic Nicole Seymour (Chapter 10, p. 191, this volume) has conceptualized "trans ecology" as sprouting off of her intriguing concept "organic transgenderism." Transecology is rooted in discussions of queer ecology except that it shifts its focus from the stronger focus on sexuality in queer studies toward questions of gender (p. 191).[3] It allows one to see the intimate intertwinement of queer and trans body politics with ecological questions and helps conceive a mutual relationship between the violent degradation of nature, its unacknowledged agency with the different forms of violence to which trans people are daily exposed (Gaard 1997; Seymour 2013, 2016; Straube 2019a; see also Bedford (Introduction, this volume); Seymour (Chapter 10, this volume). For trans politics, and trans studies in particular, discussing questions of nature in relation to trans embodiment is currently emerging and also extremely timely. As Oliver Bendorf describes in his definition of nature in the first *TSQ* issue, "Nature matters for transgender studies because of how we map (and are mapped) along boundaries of inside and out, natural and unnatural" (Bendorf 2014, 136). With this discussion on transecological aesthetics in the film *Nånting*, I follow his suggestion that transgender studies help put forward a discussion on nature that allows one to engage with questions of nature, "naturalness," and "purity" (Bendorf 2014). As biologist and philosopher Donna Haraway already stressed many years ago, nature is never "pure"—it is always inextricably layered with culture—which resulted in her conceptual creation of the term "naturecultures" (Haraway 2003, 25). Hence, transecology is central to my discussion in this chapter, as it captures the corporeal as well as discoursive entanglement of nature with trans and queer bodies.

To contextualize the discussion of the film's transecological aestheticization of nature, waste, and of the posthuman ecological intimacy of different affective materialities further, I introduce the film and the production context before developing the discussion and analysis through close readings of two sequences in the film.

Materialities of the film #1: synopsis and background

As the title of the film already suggests, "something must break" (in Swedish: nånting måste gå sönder). The main character (played by Saga Becker) refers to herself in the beginning of the film as Sebastian—as someone who is not quite Ellie yet—when she says in the first minutes of the film, "*It's as if I'm destroying myself. To become her. My dream sister: Ellie.... Her name must fill my legs. Every step I take must be hers*" (Figure 3.3). Throughout the film, Ellie suffers deeply from transphobic problems in her relationship with the second main character Andreas (played by Iggy Malmborg). Andreas is unwilling to

Figure 3.3 The main character Ellie (Saga Becker) after her shift in the warehouse.

Source: Still from *Nånting måste gå sönder*, Ester Martin Bergsmark (director), Minka Jakerson (cinematographer), Ester Martin Bergsmark/Eli Levén (script). Copyright: Garagefilm International AB.

accept her gender nonconformity which leads Ellie ultimately to break up with him. Things break in this film—on the one hand it's the relationship, but most importantly it is Ellie's "breaking out" of oppressive gender norms, which Ellie is unable to accommodate.

Andreas and Ellie are both Swedes, white, in their mid-twenties (Figure 3.4). Whereas Ellie is from the beginning of the film presented at her workplace restocking and packing at the loading dock of a warehouse, where she tries to hide her long hair under a woolen hat and retreats from awkward conversations with her co-workers, Andreas has a middle-class background. His neat-looking, hipster friends seem to situate him among class privileges Ellie doesn't have. Throughout the film, Ellie's love for Andreas is hardly ever reciprocated, or only ambivalently and temporarily, which makes it even more difficult for Ellie to let go of it. The film closes with a scene where she finally breaks up with him and finds refuge in her favorite spot in the suburbs, a hill in southern Stockholm, Högdalstoppen, where she sits on a bench overlooking the city.

Overall, the film centralizes Ellie's emancipation from her unaccepting boyfriend. The film follows the couple from their first encounter, then their first date and sex, eventually to their tragic breakup a few weeks later. From the beginning Andreas is bewildered by Ellie's femininity, confused by what this might make of him, whether he is gay if he likes to have sex with her. While Andreas himself looks relatively queer and unconventional, he also seems to be unable to reclassify himself or others without experiencing a fundamental crisis about his

Figure 3.4 Ellie and her lover Andreas in a neglected, overgrown amusement park in Stockholm during their first date.

Source: Still from *Nånting måste gå sönder*, Ester Martin Bergsmark (director), Minka Jakerson (cinematographer), Ester Martin Bergsmark/Eli Levén (script). Copyright: Garagefilm International AB.

identity. His style seems to frame him as an androgynous punk with blondish, tangled hair, a single dangling earring, eyeliner, a ragged semi-white tank top, tight black pants. His one-room flat is messy, littered with cigarette butts and empty beer cans. Despite his anarchist appearance, the boundaries of his "alternative" attitude are reached surprisingly quickly when he fears leaving the expected realm of cisheteronormativity that, as his straight middle-class-looking friends in the film emphasize, still determines his life.

Regarding the character Ellie, I made the choice to not use both names "Ellie/Sebastian" as it is written in the script and also to use female pronouns. In the film, Ellie is (still) addressed with male pronouns but they no longer seem to fit her. She hasn't yet made a suggestion for a new pronoun though. Ellie explains shortly after the opening of the film in the film's poetic voice-over that she needs to become Ellie and that she must begin to acknowledge this in herself. After the first half of the film she begins to introduce herself as Ellie to other people. I will acknowledge this direction of her transition by calling her Ellie. However, I struggle with the pronoun choice because I could also use nonbinary pronouns. "They/them/theirs" are my own pronouns. Yet "she/her" for Ellie seems more appropriate. In the voice-over Ellie says that Ellie is her "dream sister"—the one she wants to become finally—and in that voice-over she refers to Ellie with female pronouns. So, for now I choose "she/her"; maybe at another point I will choose "they/them" for Ellie.

The film *Nånting måste gå sönder* is director Ester Martin Bergsmark's debut film. A feature-length fiction film. It is the first European film by a trans/nonbinary filmmaker team and cast to be widely screened and recognized with film awards. Saga Becker, the main actress, is also the first trans person in Sweden to have won the national award for Best Female Actress, the Guldbagge award, in 2015. The script is written by Ester Martin Bergsmark and their friend and author Eli Levén. Eli Levén's semi-autobiographical book *Du är rötterna som sover vid mina fötter och håller jorden på plats* is the original foundation of the film's script. The close friendship between Ester and Eli and their earlier coming-of-age together as gender-nonconforming youth in Sweden has been beautifully explored in Ester's earlier film, the experimental documentary feature *Pojktanten* (2012). Within the genre of Trans Cinema, *Nånting* is unique in placing the often-dominant trope of gender trouble not within the trans character but in the supporting, non-trans character Andreas.

Materialities of the film #2: intimacies

The first time I saw *Nånting* was on a DVD. It arrived by mail after I had swapped it with Ester for my recently published PhD thesis on Trans Cinema. We knew each other only briefly through a mutual friend. Upon my first viewing, and despite the fact of the reduced cinematic intensity of a DVD experience, I was deeply moved by the storytelling and particularly the chosen aesthetics of this film—its rebellious eco-politics, the anarchist punk aesthetics, the extraordinary environmental irony which in all ways strays far from any moralizing dogma. The film establishes a strong intimacy that involves its entrants,[4] the cinematic figures and film's materialities: beautiful plastic bags in a forest, two human characters newly in love swimming in a dirty pond at the side of a highway, tree leaves animated by wind, spiders, ticks, a dog, human characters. Emphasis on the sound, for instance, the gushing of tree leaves in the wind, the strongly significant soundtrack, or the poetic voice-overs reflecting Ellie's thoughts, form an intensified experience of the film. The grainy close-ups, the hand-held camera, and as mentioned above, the general focus on sound were all part of the cinematic forms that drew me toward this film. Echoing multisensorial approaches to film (Marks 2000; Sobchack 2004; Straube 2014), the film in general produces an affective complexity that allows the entrant to perceive the multisensorial politics of the film as an intimate invitation into posthuman relationality. Often the film plays with an auditory incongruence: what is said is not what is heard (e.g., the highway noise that is overpowered by the rain falling on leaves). This leads me to think of not fitting in, not matching expectations. It is beautiful in the way it shifts and creates an affective proximity through the emphasis on sound: the soundtrack, the sound-scapes as well as the poetics of Ellie's voice-over are those sounds that disorient but don't distance me as an entrant. Instead they pull me in, intrigued by this new challenge to my senses and the unexpected sensuality of the film's materializing language. The sonic is connected to the visuals of plants. Close-ups of leaves and weeds initiate an intimacy between Ellie and the entrant as well as between Ellie and her ambivalent relation to nature.

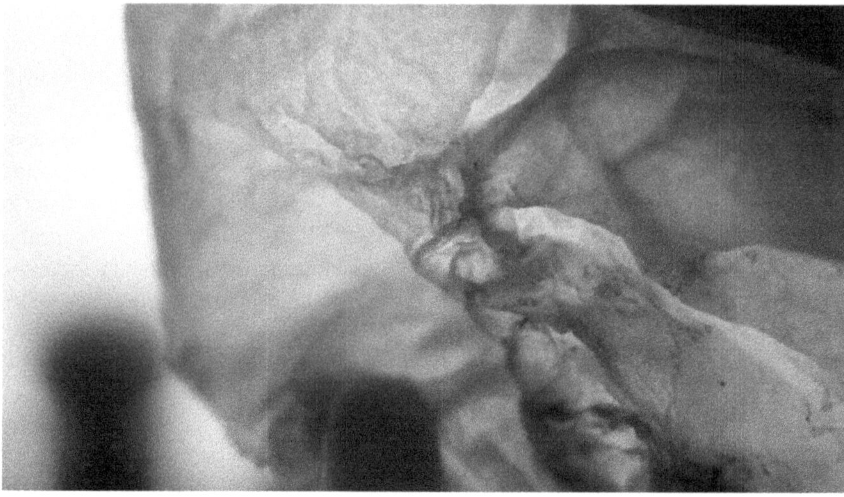

Figure 3.5 A bloodied, dried-up handkerchief which Ellie had secretly saved after Andreas had defended her against an attacker and had wiped his bleeding nose with it.

Source: Still from *Nånting måste gå sönder*, Ester Martin Bergsmark (director), Minka Jakerson (cinematographer), Ester Martin Bergsmark/Eli Levén (script). Copyright: Garagefilm International AB.

Starting with these words and affective enactments, the film projects intimacy on this level of film engagement via formal means.

Aesthetics of contamination

Repeatedly in this film, waste is enacted as a highly aesthetic object comprising close-ups of, for instance, a bloodied, dried-up handkerchief (Figure 3.5), worn underwear on the floor, shoe marks in the empty hallway, crumbs on the floor, a spider crawling over skin. This is increased by the general messiness of spaces, such as Andreas' untidy room with its scattered empty beer cans and cigarette butts everywhere, old torn clothes, smoke, blood, and even vomit and urine are among the "abject materialities" that form a polluted assemblage in this film. All these "things" (Bennett 2010, x), usually perceived as dirty, ugly, shameful even, appear as pathways into alternative sensualities—a hint at stories of livability, possibly other forms of living, and different engagements with the world. These things make up the ecology of the film and its understanding of "nature." Nature is an adventurous space, dangerous as much as liberating in this film. It is also Ellie's space more than Andreas'; a reclaimed space in which her own queer trans self meets a likeness in the unkept hill area outside Stockholm which she claims as her refuge, the garbage incineration facility right next to it, together with the highway ramps, small forest patches with dirty ponds, and generally the absence of orderly and clean surroundings.

The hill, Högdalstoppen, is partly a cruising area. She goes there to have sex with the men she meets there. It is one of those queer, impure spaces of the film that is captured as a sensual space of sexual encounter as much as a space of sensual encounter with the suburban hills themselves for Ellie, its vegetation in the close-ups, the paths that in unintentional ways map the hill and her paths through it. It is nature in an unruly space of sexual deviancy amid an otherwise clean and highly regulated metropolis. The suburb appears as the central space in this film–a space that is surprisingly rarely discussed in queer and trans research on the urban and/or the rural (Halberstam 2005; Mortimer-Sandilands and Erickson 2010; see also Huang (Chapter 5, this volume), Hogan (Chapter 7, this volume), Seymour (Chapter 10, this volume), and Thorsteinson and Joo (Chapter 2, this volume)). The suburban space in this film is not one of developed, upmarket residential areas but of semi-industrial and semi-wasteland areas, structured by highway ramps, forgotten patches of forest, remote shut-down amusement parks, active constructions sites, and mall parking lots; an urbanized but partly undisciplined wilderness. The suburbs might also be the realm that offers a refuge to those that haven't been "taken from view": the queers, the poor, the homeless, and garbage itself. Moreover, they are where people create homes that are not part of the city's regulated territory: refugees' temporary campsites, homeless people's tents, cruising areas. Ellie's gender-nonconforming self seems to be most at home in these non-categorizable spaces of the metropolitan suburbs.

In the following pages I will discuss two central film sequences of *Nånting*. In the first sequence, placed half-way into the film, he two main characters go for a swim in a small forest lake during the day of their first date. The second sequence presents the closing scene of the film after the couple's breakup in which Ellie finds a refuge atop Högdalstoppen.

The pond, murky water

Ellie and Andreas take a swim in an unappealing pond. The water seems murky, like a dirty creek at the side of a road, not water I would necessarily feel like swimming in. They had sex for the first time after a night out roaming the city. Without much sleep they have left the house again and end up sitting on the side of a mall parking lot.

"*You want to go for a swim?*" Ellie suggests. She then leads the way across the highway ramp alongside the parking lot, and into a small, quiet forest patch. The camera stays far behind; as small figures in the distance, they disappear into the undergrowth. A path leads into the woods. The forest sounds take over the scene and before even seeing them inside the forest the enhanced sound of wind in the trees almost completely overrides the previous traffic sounds that had accompanied their stop on the parking lot. The change in acoustics incites a sense of calmness and of relaxation, invoking freshness and proximity to nature. Despite the distant figures in the image that vanish between the trees, nothing is as close as the sound of the wind. This unintermediated sound remains intense

Figure 3.6 Ellie and Andreas taking off their clothes in the rain.

Source: Still from *Nånting måste gå sönder*, Ester Martin Bergsmark (director), Lisabi Fridell (cinematographer), Ester Martin Bergsmark/Eli Levén (script). Copyright: Garagefilm International AB.

also in the next scene as they walk into the forest. The camera is in close-up on their backs and heads, their feet crunching branches on the ground, birds chirping, enormous ferns line the sides of the path and brush their legs and arms. Having reached the water they undress, partly hidden from the camera among the undergrowth (Figure 3.6). It has started to rain.

The pond is surrounded by various plants, tree branches, old car tires, rocks, and a plastic bag half hanging in the water from a toppled juniper tree. A concrete wall shadows the scene in the background which is a massive cement channel guiding a thin current pipe underneath the highway and into the pond. The road itself remains nearly invisible behind the trees. Although this scene takes place close to a busy highway, there is no traffic noise—only the raindrops on the water and the trees are acoustically enhanced and the street noise is faded out the moment they enter the water. An instrumental soundtrack accompanies the scene, reserved but growing in volume as the sequence proceeds until not only the highway but also the sounds of nature have faded out. Ellie and Andreas, half-immersed, splash in the water, playful, moving closer until eventually they kiss. The scene, even with the intensity between the two main characters, stages also an iconic re-enactment of a conventional heterosexual romance scene typical in Western cinema, framed through a colonial imaginary of white sandy beaches—except here the heterosexual, cis, and white normativity is undermined and mocked by the dirty pond next to a highway and the queer couple splashing in murky water (Figure 3.7).

Figure 3.7 Ellie and Andreas bathing in the pond next to the highway.

Source: Still from *Nånting måste gå sönder*, Ester Martin Bergsmark (director), Lisabi Fridell (cinematographer), Ester Martin Bergsmark/Eli Levén (script). Copyright: Garagefilm International AB.

Despite its quiet irony, this scene contains tenderness. Andreas seems to finally stop resisting Ellie. Their intimacy is fragile. But it is also an intimacy that expand beyond their human bodies toward the polluted water, the garbage around the pond, and the proximity to the road and its noise. It is a moment of subversion of norms of purification and containment, as much as of wilderness and nature as the space of heteronormativity and masculinity (Mortimer-Sandilands and Erickson 2010). Ellie's body in particular, in the permeability of water, becomes synonymous with the understanding of non-normative bodies as "leaky bodies," as outlined by philosopher and phenomenologist Margrit Shildrick (1997). In opposition to the modernist norms of an autonomous, impermeable Cartesian ideal of the masculine, rational self–the trans body along with other non-normative bodies is always stigmatized as an "uncontained" body—one that spills and is threatening to contaminate those that come in contact with it. Ellie's body in its contact with the dirty water, but also the rain, the plants, the trash around her, seems to exemplify her as becoming "posthuman" a being entangled deeply with its intimate Others, the nature, the pollution, the animals around her, and other human, organic, and inorganic matter that surrounds her and is becoming part of her in that moment.

Uncontainable waste

Waste is "a state of constant becoming in relation" (Mehrabi 2016, 192). In this film waste troubles any notion of boundary distinction and of a polar opposition between polluted and pure; its uncontainability implodes clear distinctions between those oppositions. Waste itself is uncontainable (Radomska 2016), indeterminable (Hird 2012), and generally disturbs order, as anthropologist Mary Douglas (2002) has summarized. Waste in this film is highlighted in its relationality—in how it seeps into the world, toward human and other bodies, and how it relates and resonates with the "impure" human subjects, the two main characters, and in particular with Ellie. The film shows a tender sensuality in waste when found in spaces in which it doesn't seem to belong, where it becomes ambivalent, wanted yet unwanted, and ultimately a tactile object that seems to generate a life of its own outside its predestined arrangement (Edensor 2005). Waste objects are not only purposeless, dead objects but, in their sensual affectivity, "vibrant" (Bennett 2010) or "animate" (Chen 2012). By investigating the possibility of thinking of a plastic bag not as a discarded object but as a vibrant "thing-materiality" that is lively and potentially dangerous matter, the entanglement and the notion of intimacy with the human and more-than-human world becomes politically pressing (Bennett 2010, x).

I notice about myself that the trash in this film triggers my desire for images of a cleaner environment; it also disturbs my own romantic idea of Swedish nature. But the film's ironic approach toward trash presents a space for an ecological agenda that turns this desire upside down and dismisses the notion of purity often embedded in this longing for a clean nature, mirrored in many forms of environmental politics and its problematic repercussions for marginalized bodies (Ah-King and Hayward 2013; Di Chiro 2010; Chen 2012; Schaffer 2015). Interestingly, while the trans body in mainstream eco-politics is discussed as an unhealthy result of POP (persistent organic pollution, which includes endocrine-disrupting toxins), this scene links intensely to a contemporary political agenda. The political acuteness of such assertions shows how anti-toxicity environmentalism evokes a backlash for trans and queer politics. For instance, an *Advocate* article from early 2018 quotes a UK-based radical vegan group who claim, "transgender is an environmental developmental disorder" (Sobel 2018). It exemplifies how, in the common understanding, trans bodies are no longer only "unnatural" but clearly explicitly "contaminated" bodies. Giovanna Di Chiro critiques this and similar mainstream environmentalist rhetoric for its approach to the body as natural and pure, meaning also heterosexual and cisgender, altered only by the toxic chemical pollution that influences this otherwise "natural" body, its hormones, and its reproductive system (Di Chiro 2010). Herself part of the anti-toxin environmental justice movement, she is highly skeptical of those arguments, as they are charged with transphobic implications and materially affect trans and intersex people's lives (2010). This arm of the anti-toxin movement, she stresses, appeals "to pre-existing cultural norms of gender balance, normal sexual reproduction, and the balance of nature" (2010, 224). It is based on cisheteronormativity and the understanding of the sex/gender system as materially and discursively dualistic.

One of the radical claims I read as evoked by this film is that Ellie's intimacy with multiple contaminating objects makes her comprehensible as "polluted" herself—a polluted understanding of the body that is a reclaiming rather than a stigma. A contestation of normalizing politics that might counter such accusations of a trans body as a result of environmental pollution. Ellie is not aspiring to be a "clean," normalized subject. Instead she locates herself within this polluting assemblage that critiques the implicated reproductive hetero- and cisnormative registers of such claims and turns against what feminist biologist Malin Ah-King and philosopher Eva Hayward have critiqued as "politics of purity" (Ah-King and Hayward 2013, 2). Apart from the environmental claims, the "transsexual body," argued by Donna Haraway, has always been seen as "polluting" along with a range of other socially outcast bodies that contaminate blood lines and the vision of a purified nation, nature, and culture (Haraway 1997, 80). Consequently, this scene with the two characters bathing in the pond presents a scene of rebellion not only against the charge of the trans body as a polluted as much as polluting body but also as a critique of normalizing politics. It is rebellion against homonormative and transnormative politics that claim trans and homo as effortlessly integratable into the neoliberal discourses of normative assimilation, self-normalization, and self-optimization (Duggan 2003; Haritaworn 2014). Instead, in its transecological aesthetics the film voices a posthuman intimacy with the Other which moves beyond the dualism of polluted versus pure. This scene, moreover, draws upon a notion of intimacy that exceeds the intimacy between human bodies. Instead it allows a reformulation of intimacy as posthuman and ecological in the engagement between the human bodies, particularly the trans character's body, with waste and other "impure," intimate Others.

Posthuman ecological intimacy

The notion of intimacy in this chapter is inspired by Stacy Alaimo's concept of trans-corporeality (2008). Similar to Hayward's becoming "human with many" (2008, 70), the intimacy here is one that draws upon the inseparability of the human body from "nature" and "environment" and anything else that is more-than-human (Alaimo 2008, 238; Alaimo and Hekman 2008, 14). Trans-corporeality stresses a "movement across bodies" (Alaimo 2008, 238; 2010, 2), or a crossing of sites and the establishment of a place in which body and nature, materiality, and cultural construction become connected (Alaimo 2008, 238). It is a "contact zone," as Alaimo writes, between human and more-than-human nature (238). Trans-corporeality is the backbone of my approach toward intimacy yet not identical with it. While trans-corporeality is significant as such a "contact zone" between human and more-than-human nature, the concept of intimacy more strongly emphasizes particular forms of sensoriality, sensuality, and affectivity, which are central in this film. Alaimo refrains, however, from embedding the link between trans-corporeality and trans embodiment. To consider this extension, trans as a trans-corporeal embodiment could even further trouble boundary politics, trouble the differentiations between humans and animals, different organic matter, and its co-relation with inorganic materialities. According to Eva Hayward, trans as a wider term disturbs purification practices of space, species, and connection (Hayward 2008, 69). Trans performs a trans-corporeal "movement across bodies" (Alaimo 2010, 2), intimately entangling itself in the rebellion against boundary politics upholding the restraining modernist dichotomies of male and female, clean and dirty, animal and human, and more.

Intimacy as a scholarly concept has been rendered in various forms, yet mostly with a human-centered emphasis on sexuality, partnership, and family life which has also been queered by a range of scholars (Berlant 1998; Warner 2005). In this chapter I would like to step away from a human-centered notion of intimacy into a more open idea of intimacy as encounters, affinities, proximities, and ethical relations. Then, applying this term outside a human-centered taxonomy allows intimacy to become an essential part of the understanding of how different organic and inorganic materialities co-relate and mutually foster each other's emergence: a plastic bag at the side of the road, an unclean pond, animals, human bodies, garbage, plants, stones, and soil. Such a notion of intimacy troubles an anthropocentric form which includes but also expands the definition beyond human sexuality or other intimate human–human relations. Intimacy is thus far rarely discussed in the field of environmental humanities. Timothy Morton (2010, 2012) elaborates on ecological intimacy as constitutive for his definition of queer ecology. By addressing it via Donna Haraway's companion species concept, he explains intimacy as the interrelatedness and connection of and with other life forms (Morton 2010, 277; see also the discussion by Thorsteinson and Joo (Chapter 2, this volume)). When speaking of queer ecology and ecological intimacy, he stresses that it is centered on a "polymorphously perverse belonging (and longing) that doesn't fit in a straight box" (Morton 2010, 277). Morton further argues that such "intimacy necessitates thinking and practicing weakness rather than mastery, fragmentariness rather than holism, and deconstructive tentativeness rather than aggressive assertion" (2010, 278). Mel Y. Chen (2012) addresses intimacy via the improper intimacies in the possible transgression of boundaries between humans, animals, plants and non-living things. Similarly, the special issue "Thinking Linking" (2017) edited by Eliza Steinbock, Marianna Szczygielska, and Anthony Wagner discusses the ordering of intimacies as "sexual, communal or by species" in how they are rendered by affects and generally discuss human–animal intimacies and their conceptualizations in the field of trans studies (Steinbock, Szczygielska, and Wagner 2017, 1–2). These listed volumes are the few publications that have begun to engage with the question of intimacy in a posthuman, post-androcentric form and are extremely valuable for a definition of intimacy in this chapter.

To continue my definition of intimacies in this chapter, I turn to ecology. Ecologies, according to Catriona Mortimer-Sandiland and Bruce Erickson, are "issues of nature and environment" (2010, 5), or, in the singular, a "natural space" (2010, 6). For Morton, ecology is a "*mesh* … between species, between the living and the nonliving, between organism and environment" (2010, 275–6). In the *Oxford Dictionary of Reference: A Dictionary of Ecology*, "ecology" is defined as a "scientific study of the interrelationships among organisms and between organisms, and between them and all aspects, living and non-living, of their environment" (2010, 126). In my writing, ecology encompasses the spaciotemporality of plants, humans, animals, and inorganic materialities, and defines the interrelationships between species as well as living and non-living materalities. It spills into the notion of the posthuman while emphasizing "nature" rather than the "human" as its (possibly improper) object. To outline my approach toward the notion of the posthuman, I follow the feminist agenda of a "non-anthropocentric ontology and ethics that consider the human as necessarily enmeshed in a multiplicity of relations with human and nonhuman others" (Radomska 2016, 16). A posthuman approach is in itself closely linked to environmental politics and their shared approach to bodily materiality (Birke et al. 2004; Alaimo and Hekman 2008; Åsberg 2011, 2013), and "corpomaterial agencies and their transformative power" (Górska 2016, 24). Following Jane Bennett, this new conceptualization of corporeality compared to a humanist ontology engenders an altered understanding of feminist ethics and bodily matter. Posthumanism refuses to take the difference between human and non-human corporeality as given and challenges these categories as fixed or inherently separate (Alaimo 2010; Åsberg et al. 2011). Posthumanities and feminist materialism (Alaimo and Hekman 2008) affirm corporeality, matter, and its complex relationship to nature without the negative tendencies of over-romanticizing nature or female embodiment as in some ecofeminist traditions and without biological essentialization (Oppermann 2013).

Finally, to conclude this section, posthuman ecological intimacy is strongly connected to the enactment of waste and its proximity to the trans body. In addition, the space in which these encounters between the human characters, nature, and waste take place in this film is significant: the suburban space. A space out of place, set apart from easy classification, just as waste itself is matter out of place (Douglas 2002). The coinciding of the trans body with this suburban space is itself remarkable and offers an intriguing intimacy outside of the social orders of the urban center—an intimacy that forms an assemblage beyond the human body with polluted waters, wild growing weeds, the noise of an urban center, and the buzzle of wind in trees. It is a trans feminine character which has become "permeable" to its environment, to the nature around, and has found support in this multi-species assemblage of intimate Others.

The hill, a pile of rubbish

Throughout the film, Andreas is torn between his feelings for Ellie and his fear of the effects of his desire for her. Ellie suffers through his repeated rejections until she eventually has the strength to break up. The final sequence of the film captures this cut. It occurs after a party to which Andreas had invited Ellie the few days prior to it when they had a date. Together with his invitation he had demanded for her to tone down her femininity: "Go easy on the girly stuff!" Ellie is shocked by his suggestion. It's the final blow. He tries to apologize but she refuses to accept it. They don't have contact for a few days and then Ellie heads to the party in a beautiful new dress. When she enters the flat, she sees Andreas singing Karaoke and some of his friends watching or singing along. Taking an empty bottle from the floor, Ellie walks over to one of his friends, who had on an earlier occasion harassed her for being trans and smashes the bottle on his head. She leaves, followed by Andreas running after her, begging her not to leave him. Standing on the quiet street in the city center they talk: "I love you, Ellie. Don't go ...," both heartbroken in their remaining feelings for one another. Before she finally leaves, they kiss one last time. In the symbolic and strong scene that follows, Ellie ends up sitting on Högdalstoppen, witnessing the sun rise at the city's horizon.

She sits on a bench on the top of the hill, which she had been visiting often in the past weeks. It is still almost dark, mild light coming up in the night sky. A middle-aged woman with her dog appears hiking up the steep hill, panting from the slope (Figure 3.8). While calling for her little dog Tage, she becomes aware of Ellie and her heartbroken state. First to her dog and then to Ellie, she says, "Phew, what a hill! We did well, Tage! ... You're not afraid of dogs, I hope? Tage's a nice dog ... What a pretty dress. It suits you." Ellie responds with a weak but warm smile. The woman, while enjoying the view, continues: "All we're standing on ... is made of waste. We are standing on a pile of rubbish!" Having emphasized this, there seems to be nothing more to say. They smile at each other one last time (Figures 3.9 and 3.10). Then the woman and her dog turn away and continue their walk.

Figure 3.8 Ellie grieving over her breakup from Andreas on the hill Högdalstoppen (see Figure 3.1).

Source: Still from *Nånting måste gå sönder*, Ester Martin Bergsmark (director), Minka Jakerson (cinematographer), Ester Martin Bergsmark/Eli Levén (script). Copyright: Garagefilm International AB.

Figure 3.9 Ellie meeting another human being on the hill and listening to her comforting words.

Source: Still from *Nånting måste gå sönder*, Ester Martin Bergsmark (director), Minka Jakerson (cinematographer), Ester Martin Bergsmark/Eli Levén (script). Copyright: Garagefilm International AB.

Figure 3.10 The woman who walks her dog shares her knowledge about Högdalstoppen with Ellie.

Source: Still from *Nånting måste gå sönder*, Ester Martin Bergsmark (director), Minka Jakerson (cinematographer), Ester Martin Bergsmark/Eli Levén (script). Copyright: Garagefilm International AB.

A hill of rubbish—rubble, in fact. The houses torn down during Stockholm's 1950s modernization project were collected there to be developed into an urban park, a meeting place in "nature" in the vicinity of the city, designed by urban architect Holger Blom. It was abandoned before it was completed. As a failed project of modernity, it is now a cruising area and unattended space used by the people of Stockholm for walks and picnics. It is lined by paths that lead nowhere, that end abruptly as reminders of the abandoned park project; new makeshift trails extend them. The purposelessness of the constructed tracks is an affront to late capitalist aspirations for productivity and efficiency. This area in Högdalen, "Högdalstoppen Nr. 1"—one of three hills—is now a rare, semi-urban wilderness uncommon in contemporary Western metropoles, where usually every empty, open space is quickly sold off and developed. In this former park project, the hill now holds a vague memory of its originally planned success and its failure in achieving to become this productive, domesticated place.

On this hill, waste no longer consists of small things; instead it has become a solid, enormous site, a landmark. The intimacy of things that earlier accompanied Ellie's bath in the pond is carried on in this scene and manifested even more strongly in the force of the hill itself, its strength, its integration into the life forms of the humans, animals, and plants at Stockholm's outskirts. The hill becomes Ellie's final companion in the film. It had also been her first in the opening scene of the film. Symbolizing the fissures of a capitalist society in its failure to become a respectable city park, it is also a place symbolizing the expectations demanded of her by Andreas that had carried her nearly to a breaking point. It might also speak of the continuous call of a normative, judgmental society to normalize herself to fit into a pre-described, expected category. Ellie at this point has struggled at the most intimate level with these demands upon her. Her outlook on that hill now is hopeful with the view on the horizon, occupying the place that seemingly divides nature (hill and forest) and culture (city) and implicitly references its intertwinement, its undivided natureculture.

The hill becomes the companion to Elllie's own socially abject self. As Ester, the director of the film, said to me in a conversation, the hills are the "dirty outside in a world that otherwise looks so clean and functional."[5] For Ester, these hills are "spaces to breathe, and to build other types of life." They are spaces of possibility.

Hence, it is not only the hill, its rubble, and plants that sustain Ellie in this moment but also the woman she meets there and her animal companion that speak of their own particular posthuman ecological intimacies. As a woman of color, the dog-walker is, similarly to Ellie's trans body, conventionally not fully included in the category of the human. Both the trans and the Black body are bodies with a history of dehumanization (Weil 2017). A dog, as the sole non-human animal in this scene, is rendered as the ultimate living being without agency or recognition. The animal in this scene seems to echo the lack of agency and the silence imposed upon those in society who are unable to accommodate social norms (Figure 3.11).

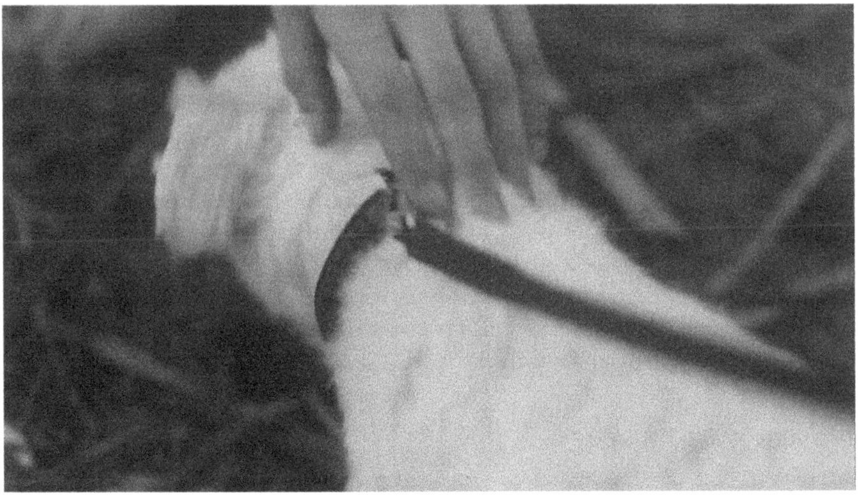

Figure 3.11 The dog Tage as one of the parts in the posthuman intimate assemblage of humans, animals, and nature.

Source: Still from *Nånting måste gå sönder*, Ester Martin Bergsmark (director), Minka Jakerson (cinematographer), Ester Martin Bergsmark/Eli Levén (script). Copyright: Garagefilm International AB.

These three living human and nonhuman beings represent in the film those joined by a loss of "purity" and uncontainability in a Western, modernist understanding of embodiment (Chen 2012, 7). Their temporary assemblage forms a posthuman ecological intimacy alongside the hill and the collapsed, wasted houses underground. In this assemblage they undermine the meaning of "natural," appropriate impurity as a site of strength, and create a utopian trajectory of "unnatural" nature. The waste, the trans body, the non-white body, the dog, and the flora and fauna of this hill become agential in this lost project of modernity and the keepers of a new space-time for posthuman affinities.

Space and the experience of space is then part of becoming trans, or of being queer or "othered" multiple ways. Ellie's investment in her location as something that determines her and elaborates the significance of space is related to a sense of community and self. Generally, space and becoming through space interacts with every aspect of being a human being and particularly with community, belonging, and identity (Bremer 2011; Siverskog 2016). Space and sense of self are intimately entangled and, for instance, have a strong significance in social justice movements and the contestation of space as a livable space (Gieseking 2014). Those who are socially marginalized are not only on the social periphery but are always also structurally pushed to the outskirts of cities due to factors such as unregulated housing markets, gentrification, and structural poverty, and they are often in closer proximities with environmental pollution and toxins. Things, human beings, animals, and intimate wasteful Others are thus a result of spatialization as much as agents in forming urban, suburban, and rural spaces themselves. Hence, the outskirts in this film have been appropriated as zones of livability for those that do not belong, that are excessive, contaminating, and wasteful. They are the places of those that do not belong to the white, straight, middle-class norm. Ellie intimately assembles herself alongside plants, water, and green hills that, despite the usual cultural interpretations as being "lively" and "clean," are slightly treacherous, ambiguous, dirty geographies that point toward possibility and livable futures. Their intimacy is posthuman as well as ecological—moved to the margins by poverty and a racist, speciesist, and cisheteronormative paradigm of productivity, whiteness, and health. They have together become matter out of place, but they also *matter* places differently themselves.

Intimate wasteful others

In this chapter and particularly in the two discussed film sequences, I have engaged with the film's rebellious reappropriation of contamination. This unruly aesthetics of waste and pollution folds into a force that nurtures a transecological aesthetic and which speaks to questions of livability and co-becoming with intimate Others. As the cisnormatively unwanted, wasteful body, the trans body seeks a place along the social margins in this film. Taken from view, from being seen as the one that does not fit into parameters of binary gender code. This body spills, it is excessive, it is ultimately uncontainable (Shildrick 1997, 2002) while it blends with the polluted materialities of its environment. The repercussions of this "leaky" body are fierce and much feared in a society that wants to keep itself clean, pure, and categorizable. The polluting assemblages enacted in this film are an assembly of intimate Others. They turn the ecological "leave no trace" policy upside down and make themselves a home in a new place, in nature. The hill, the pond, the outskirts, the aesthetics of garbage all around are no longer the dirty "outside" but become spaces of livability for those that are outcast on the basis of their status as wasteful, uncontained bodies. Finally, the posthuman ecological intimacy that unfolds in *Nånting måste gå söder* is one that engages with a body that is becoming with many, as entangled with multiple others. Understanding the human body's proximity with other human bodies, animals, plants, organic and inorganic matter as intimate, based on affectivity, new affinities, is anchored in the knowledge that nature is our place, our only home.

Notes

1 I thank Nina Lykke, Olga Cielemęcka, and Tara Mehrabi for their suggestions and help in sharpening the focus of this chapter. I am grateful to Annika Ruth Persson and Joe Castrovinci for their editorial work on the text, as well as Magdalena Górska, Marinette Grimbeek, and Ida Linander for references and comments. I am especially

thankful to Ester Martin Bergsmark for our shared conversations and for making beautiful films.

2 Transgender Europe offers valuable information on forced sterilization practices in different European countries: http://tgeu.org/24-countries-in-europe-still-require-sterilization-from-trans-people/. Denmark discontinued forced sterilization in 2014, Sweden in 2013, and the German law was abolished in 2011. Swedish trans activists ensured financial compensation in April 2016; see RFSL: www.rfsl.se/en/organisation/tvaangssteriliseringar-och-skadestaand/ekonomisk-kompensation-till-tvaangssterili serade/.

3 I would like to thank Martina Böll for her presentation on life writing and her Guattarian ecosophy-inspired application of transecology during the sixth Nordic Trans Studies conference at NTNU in Norway and her paper "Being and Becomings. Ecological Perspectives on Contemporary Trans Life Writing" (September 2019).

4 I am critical of the term "viewer," as it centralizes an occularcentric experience of audio-visual media. I created the term *entrant* in order to emphasize a multisensorial engagement with film (Straube 2014).

5 Ester and I have met several times over the past few years, connected at first through their flatmate Johanna who had become a friend of mine in Berlin. Later we met at film festivals and were both invited to Linköping University to do an hour-long public Q+A after the screening of *Nånting*. The event was organized by the Forum for Gender Studies and Equality and hosted by Ulrika Engdahl. This quote was part of our conversation that evening (September 11, 2015).

References

Ah-King, Malin and Eva Hayward. 2013. "Toxic Sexes. Perverting Pollution and Queering Hormone Disruption." *O-Zone: A Journal of Object-oriented Studies* 1: 1–12.

Alaimo, Stacy. 2008. "Trans-corporeal Feminisms and the Ethical Space of Nature." In *Material Feminisms*, edited by Stacy Alaimo and Susan Hekman, 237–64. Bloomington: Indiana University Press.

Alaimo, Stacy. 2010. *Bodily Natures: Science, Environment and the Material Self.* Bloomington: Indiana University Press.

Alaimo, Stacy and Susan Hekman. 2008. "Introduction: Emerging Models of Materiality in Feminist Theory." In *Material Feminisms*, edited by Stacy Alaimo and Susan Hekman, 1–19. Bloomington: Indiana University Press.

Åsberg, Cecilia. 2013. "The Timely Ethics of Posthumanist Gender Studies." *Femininistische Studien* 1: 7–12.

Åsberg, Cecilia, Redi Koobak, and Ericka Johnson. 2011. "Beyond Humanist Imagination." *NORA—Nordic Journal of Feminist and Gender Research* 19(4): 218–30.

Bendorf, Oliver. 2014. "Nature." *TSQ: Transgender Studies Quarterly* 1(1–2): 136–7.

Bennett, Jane. 2010. *Vibrant Matter. A Political Ecology of Things*. Durham, NC: Duke University Press.

Berlant, Lauren. 1998. "Intimacy: A Special Issue." *Critical Inquiry* 24(2): 281–8.

Birke, Linda, Mette Bryld, and Nina Lykke. 2004. "Animal Performances: An Exploration of Intersections between Feminist Science Studies and Studies of Human/Animal Relationships." *Feminist Theory* 5(2): 167–83.

Bremer, Signe. 2011. *Kroppslinjer–Kön, transsexualism och kropp i berättelser om könskorrigering.* PhD Dissertation. Göteborgs universitet. Lund: Makadam.

Burke, Mary C. 2011. "Resisting Pathology: GID and the Contested Terrain of Diagnosis in the Transgender Rights Movement." *Advances in Medical Sociology* 12: 183–201.

Butler, Judith. 2004. "Undiagnosing Gender." In *Undoing Gender*, edited by Judith Butler, 274–98. New York: Routledge.

Chen, Mel Y. 2012. *Animacies: Biopolitics, Racial Mattering, and Queer Affect*. Durham, NC: Duke University Press.

Cole, July. 2016. "Lucky, Trashy: Sunshine and Other Queer Waste." Paper presented at the International Conference of the European Network of Political Ecology (ENTITLE), on the panel "Who will queer political ecology? or Cute goners, (in) human thinkers, and queer wastoids," Stockholm, March 20–23.

Di Chiro, Giovanna. 2010. "Polluted Politics? Confronting Toxic Discourse, Sex Panic and Eco-normativity." In *Queer Ecologies: Sex, Nature, Politics, Desire*, edited by Catriona Mortimer-Sandilands and Bruce Erickson, 199–230. Bloomington: Indiana University Press.

Douglas, Mary. 2002. *Purity and Danger: An Analysis of the Concept of Pollution and Taboo*. New York: Routledge.

Duggan, Lisa. 2003. *The Twilight of Equality? Neoliberalism, Cultural Politics, and the Attack on Democracy*. Boston, MA: Beacon Press.

Edensor, Tim. 2005. "Waste Matter: The Debris of Industrial Ruins and the Disordering of the Material World." *Journal of Material Culture* 10(3): 311–32.

Engel, Antke and Renate Lorenz. 2013. "Toxic Assemblages, Queer Socialities: A Dialogue of Mutual Poisoning." *E-flux journal* 44(April). Accessed August 15, 2016. www.e-flux.com/journal/toxic-assemblages-queer-socialities-a-dialogue-of-mutual-poisoning/.

Enke, Finn Anne. 2012. "Introduction: Transfeminist Perspectives." In *Transfeminist Perspectives. In and Beyond Transgender and Gender Studies*, edited by Finn Anne Enke, 1–15. Philadelphia, PA: Temple University Press.

Gaard, Greta. 1997. "Toward a Queer Ecofeminism." *Hypathia* 12(7): 114–37.

Gieseking, Jen Jack. 2014. "A Queer Geographer's Life as an Introduction to Queer Theory, Space and Time." In *Queer Geographies. Beirut, Tijuana, Copenhagen*, edited by Lasse Lau, Mirene Arsanios, et al., 14–21. Saline, MI: McNaughton and Gunn.

Górska, Magdalena. 2016. *Breathing Matters: Feminist Intersectional Politics of Vulnerability*. PhD Dissertation. Linköping: Linköping University Press.

Halberstam, Judith Jack. 2005. *In a Queer Time and Place. Transgender Bodies, Subcultural Lives*. New York: New York University Press.

Haraway, Donna. 1997. *Modest_Witness@Second_Millenium. FemaleMan©_Meets_Onco_ Mouse™. Feminism and Technoscience*. New York: Routledge.

Haraway, Donna. 2003. *The Companion Species Manifesto: Dogs, People, and Significant Otherness*, Vol. 1. Chicago, IL: Prickly Paradigm Press.

Haritaworn, Jin. 2014. *Queer Lovers and Hateful Others. Regenerating Violent Times and Places*. Chicago, IL: Chicago University Press.

Hayward, Eva. 2008. "More Lessons from a Starfish: Prefixial Flesh and Transspeciated Selves." Special issue, "Trans-," *WSQ: Women Studies Quarterly* 36(3–4): 64–85.

Hird, Myra. 2012. "Knowing Waste: Towards an Inhuman Epistemology." *Social Epistemology* 26(3–4): 453–69.

Krieg, Josephine. 2013. "A Social Model of Trans and Crip Theory. Narratives and Strategies in the Redefinition of the Pathologized Trans Subject." *lambda nordica* 3–4: 33–53.

Marks, Laura U. 2000. *The Skin of the Film: Intercultural Cinema, Embodiment, and the Senses*. Durham, NC: Duke University Press.

Mehrabi, Tara. 2016. *Making Death Matter. A Feminist Technoscience Study of Alzheimer's Sciences in the Laboratory*. PhD Dissertation. Linköping: Linköping University Press.

Mortimer-Sandilands, Catriona. 2005. "Unnatural Passions? Toward a Queer Ecology." *Invisible Culture* 9. Accessed July 1, 2016. www.rochester.edu/in_visible_culture/Issue_9/title9.html.

Mortimer-Sandilands, Catriona and Bruce Erickson. 2010. "Introduction: A Genealogy of Queer Ecologies." In *Queer Ecologies. Sex, Nature, Politics and Desire*, edited by Catriona Mortimer-Sandilands and Bruce Erickson, 1–47. Bloomington: Indiana University Press.

Morton, Timothy. 2010. "Queer Ecology." *PMLA* 125(2): 273–82. Accessed August 20, 2016. www.sussex.ac.uk/webteam/gateway/file.php?name=pmlamorton2010.pdf&site=58.

Morton, Timothy. 2012. *The Ecological Thought*. Cambridge, MA: Harvard University Press.

Oppermann, Serpil. 2013. "Feminist Ecocriticism. A Posthumanist Direction in Ecocritical Trajectory." In *International Perspectives in Feminist Ecocriticism*, edited by Greta Gaard, Simon C. Estok, and Serpil Oppermann, 19–36. New York: Routledge.

Radomska, Marietta. 2016. *Uncontainable Life: A Biophilosophy of Bioart*. PhD Dissertation. Linköping: Linköping University Press.

Schaffer, Guy. 2015. "Queering Waste Through Camp." *Discard Studies*. Accessed August 15, 2016. http://discardstudies.com/2015/02/27/queering-waste-through-camp/.

Seymour, Nicole. 2013. *Strange Natures: Futurity, Empathy, and the Queer Ecological Imagination*. Champaign: University of Illinois Press.

Seymour, Nicole. 2016. "Trans Ecology and the Transgender Road Narrative." *Oxford Handbook Online*, edited by Greg Garrard, 1–28. Oxford: Oxford University Press.

Shildrick, Margrit. 1997. *Leaky Bodies and Boundaries: Feminism, Postmodernism and (Bio)Ethics*. New York: Psychology Press.

Shildrick, Margrit. 2002. *Embodying the Monster: Encounters with the Vulnerable Self.* Thousand Oaks, CA: Sage.

Siverskog, Anna. 2016. *Queera livslopp. Att leva och åldras som lhbtq-person i en heteronormativ värld*. PhD Dissertation. Linköping: Linköping University Press.

Sobchack, Vivian. 2004. *Carnal Thoughts: Embodiment and Moving Image Culture*. Berkeley: California University Press.

Sobel, Ariel. 2018. "Vegan Group Says 'Chemical Pollutants Make People Gay and Trans'." *The Advocate*, July 19, 2018. Accessed August 1, 2018. www.advocate.com/environment/2018/7/19/vegan-group-says-chemical-pollutants-make-people-gay-and-trans.

Spade, Dean. 2006. "Mutilating Gender." In *Transgender Studies Reader 1*, edited by Susan Stryker and Steven Whittle, 315–32. New York: Routledge.

Steinbock, Eliza, Marianna Szczygielska, and Anthony Wagner. 2017. "Thinking Linking Editorial introduction." *Angelaki. Journal of the Theoretical Humanities* 22(2): 1–10.

Straube, Wibke. 2014. *Trans Cinema and Its Exit Scapes: A Transfeminist Reading of Utopian Sensibility and Gender Dissidence in Contemporary Film*. PhD Dissertation. Linköping: Linköping University Press.

Straube, Wibke. 2019a. "Toxic Bodies. Ticks, Trans Bodies, and the Ethics of Response-Ability in Art and Activist Writing." *Environmental Humanities* 11(1): 216–38.

Straube, Wibke. 2019b. "Trans Natures. Intimacies, Trans Bodies and Pollution in Environmental Art and Film." Keynotes. "Trans Realities." – Sixth Nordic Trans Studies Network Conference, September 11–13, 2019, NTNU, Trondheim, Norway.

Stryker, Susan. 2006. "My Words to Victor Frankstein above the Village of Chamounix: Performing Transgender Rage." In *Transgender Studies Reader 1*, edited by Susan Stryker and Steven Whittle, 244–56. New York: Routledge.

Stryker, Susan, Paisley Currah, and Lisa Jean Moore. 2008. "Trans-, Trans, or Transgender?" Special issue, "Trans." *Women Studies Quarterly* 36(3–4): 11–22.

Warner, Michael. 2005. *Publics and Counter Publics.* New York: Zone Books.

Weil, Abraham. 2017. "Trans*versal Animacies and the Mattering of Black Trans*Political Life." *Angelaki. Journal of the Theoretical Humanities* 22(2): 191–202.

Winters, Kelley. 2005. "Gender Dissonance: Diagnostic Reform of Gender Identity Disorder for Adults." *Journal of Psychology & Human Sexuality* 17(3): 71–89.

Woelfle-Erskine, Cleo. 2016. "My Dead Cutie: Queer Trans-species Love in Wastelands." Paper presented at the International Conference of the European Network of Political Ecology (ENTITLE), on the panel "Who will queer political ecology? or Cute goners, (in)human thinkers, and queer wastoids," KTH, Stockholm, March 20–23.

Woelfle-Erskine, Cleo. n.d. "Queer x Trans x Ecology: Towards a Field Praxis." In *Underflow: Transfiguring Riverine Ecology*. In preparation.

Filmography

Bergsmark, Ester Martin (director). 2012. *Pojktanten* [*Shemale Snails*]. Script by Ester Martin Bergsmark. Danish Upfront Films. 75 min. Sweden.

Bergsmark, Ester Martin (director). 2014. *Nånting måste gå sönder* [*Something Must Break*]. Script by Ester Martin Bergsmark and Eli Levén. Garagefilm International AB. 80 min. Sweden.

4 A journey through eco-apocalypse and gender transformations

New perspectives on Angela Carter's *The Passion of New Eve*

Julia Kuznetski

Speaking on the relationship of the environment and the body, Patrick Murphy (2013, 50) emphasizes that identity is not found in the body, since our bodies are not final, and thus individuals are "never complete, never finished." Murphy agrees with Butler (2004, 219), who suggests that although "we struggle for rights over our own bodies, the very bodies for which we struggle are not quite ever only our own," and addresses the problem that in today's world, where there is no complete bodily autonomy and a body is always formed in "the social crucible" (Butler 2004, 22), exposed "to the gaze of others but also to touch and to violence" (ibid., 21). In this vein, various processes, including sex changes, are but a part of a process of a body's becoming, but "never complete." Likewise, recent feminist theory has employed the idea of body-in-place as an alternative to both essentialism and social constructionism—a body dependent on the affirmation of others on earth (Mann 2006, quoted in Gaard 2016, 186). Greta Gaard (2016, 182) highlights the interdependence of human, animal, and environmental health narratives as a key legacy of the "feminist, anti-toxic and environmental justice movements of the twentieth century." Addressing the case of women, people of color, and representatives of the LGBT community, Gaard insists that these marginalized groups are most affected by climate change (due to poverty, vulnerability, and exclusion), and yet are underrepresented in climate talks—thus, the present-day narrative, as Gaard (2016, 180–81) rightly[1] observes, is "truncated." An alternative would be a "complete narrative, a 'body-in-place'," a transcorporeality that is a constant becoming with "earth's others" (in Plumwood's terms), involving and valuing "the intersecting differences of gender, sexuality and species with differences of race, class, ecology and nation" (Gaard 2016, 186–7), allowing us to respond more completely and meaningfully to climate change injustices.

In this chapter, I focus on environmental and gender discourse as well as transcorporeality in Angela Carter's novel *The Passion of New Eve* (1977). Analyzing how the two important agendas are addressed and intertwined, I propose a new reading of Carter, focused on the intersection of human and nature and the elusiveness, liquidity, non-finality, and interpermeability of habitual notions.

The Passion of New Eve

Angela Carter (1940–92) is not known to have views on ecology or climate change. Ironically, however, her own life was claimed by lung cancer at the age of 51—the result, as her friends saw it, of never giving up living and walking in London, despite having given up smoking for decades. Environmental catastrophe and human--nature connections, as well as the idea of rejuvenation and rebirth through contact with the primordial wilderness, are obvious in many of her writings, from the apocalyptic *Heroes and Villains* (1969) to the playful *Nights at the Circus* (1984), in which the formerly rational male protagonist is purified through amnesia and shamanism in the primordial snows of Siberia, and puts aside his gender convictions and embraces a bird woman, along with all of her biology and mythology.

The Passion of New Eve (1977, referred to hereafter as *Eve*) recounts the adventures of Evelyn, a careless young man from England, who abandons his black girlfriend in New York after an illegal abortion results in her mutilation, and then travels across the US, is captured in an Arizona desert by an Amazon community that transforms him into a woman, Eve, and is subjected to surgery, hormones, and "training" in womanhood and the inculcation of ideas of passiveness, submissiveness, and suffering. A parallel story that unfolds in the novel is that of Hollywood star Tristessa, the "perfect woman" of Evelyn's adolescent fantasies, who eventually turns out to be a drag queen. Tristessa had once decided to become a woman, but now consents to becoming a man, after Eve had become—technologically and emotionally—a woman. Their marriage, forced upon them by Zero, a homophobic and violent harem-keeper, eventually turns into love and self-knowledge, bringing about Eve's pregnancy and Tristessa's murder by a Christian radical. Finally, Eve goes through a ritualistic passage in the caves of California, and departs for the ocean, which signifies a beginning and an end.

Critics have discussed *Eve* as a dystopian feminist novel, with all the necessary ingredients present—an apocalyptic setting, a world disintegrating into solipsism, feminist radicalism, futuristic/sci-fi elements in the description of Beulah, an underground community in Arizona, as an Amazon utopia, plus elements of the grotesque and parody as a postmodern necessity. In the closing years of the twentieth century, with the proliferation of gender and queer theory, Carter again came into the limelight for her ingenious ideas on the construction of gender, which resulted in a posthumous recognition and a certain critical "after-the-fact 'Butlerification' of Carter" (Bristow and Broughton 1997, 19; Trevenna 2002), whose texts were "queer avant le lettre" (Johnson 1997). The transgender theme in Carter's oeuvre has since been touched upon by a few scholars (Johnson 1997; Caroll 2011), with conflicting conclusions regarding the authenticity of her rendering of the transsexual experience—gender transgression as textualized subjectivity (Johnson) or a manifestation of violence (Caroll). In describing the transgender turns of the plot, Peach (2009, 118) focuses not on the trans experience per se, but on the implications of identities being "imaginary and provisional rather than fixed and closed," with the biological differences between men and women being "not as important in the construction of gender identities as their elaboration in complex cultural codes" (ibid., 117). What is more, neither its ecological/ecocritical agenda nor its transecological connections have been discussed sufficiently, although the environmental agenda in *Eve* is quite obvious to a twenty-first-century ecocritic, and manifests itself in a number of environmental tropes—apocalypse and wilderness, consumption and pollution, the role of the Earth and its connection to the body, ideas of posthumanism, cyborgs, materiality, and transcorporeality, and, finally, a possibility of the rebirth of both the Earth and a body as new, transecological and interpermeating realities.

Structured superficially around Evelyn/Eve's journey across the Atlantic and America, but focused really across his/her self-identity, the novel has been described as a *Bildungsroman* (Peach 2009, 101) and a "picaresque novel" (Edwards 1998, 226). It is also important to point out that it is written in the form of an autobiography, and therefore contains a strong confessional element, recalling Jan Morris' *Conundrum* (1975), which recounts her experience with gender dysphoria and eventual sex change. According to Salamon (2010, 173), *Conundrum* established "the structural conventions of trans autobiography" as a "journey from mistaken to true sex." Published but three years before *Eve*, it employs the notion of place as a metaphor for identity, in which "gender is analogized to a country, and membership in one or the other gender … a kind of nationalism," with the author's extensive travels, "that incessant wandering" (Morris, quoted in Salamon 2010, 175) being the expression of her inner journey, or the conundrum of the title. Similarly, in Carter's novel, (1977, 144) Tristessa's self-perception as a woman in a man's body is recounted as a kind of journey: "For hours, for days, she had wandered endlessly within herself, but never met anybody, nobody." Thus, body, identity, gender, and landscape are linked in *Eve*, although we cannot speak of an "end" to the journey for the protagonist, nor a "true" gender arrived at as a result.

This linking of the body and the environment, imagining the human body as intermeshed with the material world and more-than-human others has been described by materialist ecocritic Stacy Alaimo (2010) as "transcorporeality," emerging, in the theorist's words (2010, 2), from "the literal contact between human corporeality and more-than human nature." Transcorporeality emphasizes the movement across bodies, that is, the constant change, transit and interchange between various bodies—human and nonhuman, or with the nature within nature. Nature, Alaimo stresses, cannot be separated from ourselves, because it is not part of a dualist culture/nature separation but is always "as close as one's skin" (ibid.)—meaning that our skin itself largely consists of what we think of as "others"—bacteria, waste, toxins, chemical agents, DNA traces of interaction with others—in a word, we ourselves become complex ecological systems, constantly interacting with something within and without. In Butler's (2004) words, we are always "besides ourselves," whereof the body is not understood as "static or accomplished fact" (Butler 2004, 29), but an ageing process, "a mode of becoming that, in becoming otherwise, exceeds the norm, and makes us see how realities to which we thought we were confined are not written in stone." By "exceeding the norm" Butler of course means heteronormativity, and that of gender institutionalized as "reality." If that reality is questioned, however, and substituted for a "possibility," the normalizing idea about what our bodies "ought" to be is contested. The realm of that possibility, for Butler, is transgendered[2] movements that demonstrate "transformative potential" and possibilities beyond the norm. A common argument of course is that these articulations are "against nature"—but both Alaimo's and Butler's argument is that in actuality nature *is* the realm of possibilities. Alaimo underscores that our bodies are constantly in transit anyway, with cells renewing themselves or ageing and deteriorating, or being eaten by other cells beyond our control. What is more, she recounts Hird's radical statement that "the *vast* majority of cells in the human body are intersex" (Alaimo 2010, 5, emphasis in original). Thus, in nature, nothing is static, nothing is normative, and nothing is impossible. The environmental tropes and transformations depicted in *Eve* reveal those complexities.

Environmental tropes: apocalypse and wilderness

Apocalypse and *wilderness* are listed among environmental tropes by Greg Garrard in his seminal *Ecocriticism* (2004, 184), which also lists *pollution* as catastrophic environmental destruction; *pastoral* as the spatial distinction of town and country with the temporal distinction of idyllic past and "fallen" present; *wilderness* representing purity, the sublime, and the motif of escape; and *Earth* as a super-organism, a self-regulating system with animals consisting not only of domestic and wild fauna but also the idea of cyborgs and biodiversity.

Speaking of apocalypse as the most obvious environmental trope in *Eve*, we may observe that it focuses specifically on America, which is similarly described in Eliot's *The Waste Land* as a place of sterility and disintegration. A late twentieth-century agenda (e.g., the problem of consumerism) manifests itself on Evelyn's arrival on the continent: "Welcome to the country where Mouth is King!" (10).

In describing consumerism as a mark of today's world, Patrick D. Murphy (2013, 81) insightfully stresses that consumption is a type of addictive disease, like alcohol and drug addiction. He points out how the value paradigm in the post-World War II US shifted from "worker" to "consumer" as "the label of the average adult." Likewise, for Baumann (2000, 72–3), consumption is a "compulsion turned into addiction," a "race" that this consumer society is running, with production turning into "production of consumers" (ibid., 75) and guided by a new kind of pleasure principle (ibid.). Murphy (2013, 81) attributes this to the release of chemical stimulants to the brain, making consumption addictive, both physiologically and psychologically. The latter is explained by Baumann (2013, 73) with reference to Ferguson as a link of consumption to self-expression, and to notions of taste and discrimination, whereby the individual "expresses himself or herself through their possessions," thus creating an illusion of identity, of "a coherence and a unity" (ibid., 82), which is in fact an optical illusion—not individual freedom, but "the freedom of self-identification through the use of mass-produced and merchandized commodities" (ibid., 84).

Consumption, theorized by Baumann and Murphy as a disease of our times, which has acquired a global scale, and needs, in Murphy's (2013, 81) view, urgent therapy and rehabilitation, is also a part of Carter's (1977, 10) dystopian New York, "the land of comestibles." We see the idea of consumption at its worst manifest in the scene at a night drugstore (Carter 1977, 19) where Evelyn meets, while buying cigarettes and items of "edible Americana," his short-term girlfriend Leilah, who is moving around the shop's paraphernalia on "fetishistic high heels" (ibid.). Evelyn's immediate determination to "have her"—like a packet of cigarettes or a hamburger—is as with Murphy (2013, 84) consumption as "taking": deconstructing the morphemes *–tain* in "sustain" and "contain" meaning "to hold" as antonymous to *–sume* in "consume," meaning "to take." The eminent ecocritic (ibid.) also stresses that "in English, one synonym of 'consume' is 'squander.'"

Consuming as "taking" and "squandering" bears horrible consequences in Carter's novel: in both Evelyn's relationship with Leilah and in the disintegrating state of the city itself, which mirrors that relationship. The apocalypse invading the city is obvious, visible in terror (hotel fire), the rise of extremist groups (blacks and women, "Beware Women!" (Carter 1977, 11)), competing religious missionaries and street preachers, unprecedented numbers of elderly beggars, garbage and litter in the streets, and rats. This all has a clear ecological dimension: skies are described as unnatural, synthetic, as are the people—starting with the idea of humanity preserved in celluloid in the description of Tristessa's movies to the futuristic Beullah.

That the apocalypse Evelyn witnesses in New York is first of all an eco-apocalypse is clear in the description of the landscape: a "wasted, inner-city moon to which pollution lent a mauvish tinge" (ibid., 20), while the skies

> were of strange, bright, artificial colours—acid yellow, a certain bitter orange that looked as if it would taste of metal, a dreadful, sharp, pale, mineral green—lancinating shades that made the eye wince. From these unnatural skies fell rains of gelatinous matter, reeking of decay. One day, there was a rain of, I think, sulphur.
>
> (ibid., 12)

The city is also a kind of eerie jungle, with refuse in the "vile street," rats congregating among hamburger stands (ibid., 21) and, "as if cut out of dark paper and stuck against the sky … the negative perspectives of the sky-scrapers" (ibid., 23), an "arid world of ruins and abandoned construction sites" (ibid., 21). Fleeing New York, Evelyn recognizes an intertwining of the city and his own inner world, with its manifestations of cruelty, violence, and callousness. These constitute a universal, global tendency that is impossible to escape, even by fleeing the city: "the darkness and confusion were as much my own as that of the city and I took the sickness with me since I was myself infected … as myself a carrier of the germ of a universal pandemic of despair" (ibid., 37).

Desert as an ambiguous "wilderness"

Desert constitutes the next stage in this picaresque narrative presenting, in ecocritical terms, a contrast to the previously rendered polluted and apocalyptic city. With its pureness, colorless plainness, sand, and air, "the primordial light unexhausted by eyes" (Carter 1977, 38), it can even be seen as a kind of contrasting pastoral space outside the city that the protagonist believes will cure and purify him. In his escape to the desert, Evelyn seeks to become an American Adam, fleeing civilization in the manner of Thoreau, to live free and uncommitted from ties of society and human relationships. This "uncommitment" is now very urgent for Evelyn, who is leaving behind not just a perishing city but his mutilated, sterile girlfriend, and his own "not very tender" (ibid., 36) consciousness. Buell (2005, 67) has pointed out the function of wilderness as a "therapeutic refuge" in classical American writings (Thoreau's *Walden*). Earlier (1989, 1), he described wilderness as a liminal site for an American male recoiling from "adult responsibility associated with women-dominated" society. This is exactly what Evelyn hopes to find in the wilderness—a therapeutic refuge away from responsibility—when he juxtaposes the purity of the desert with the corruption and hypocrisy of the "them" (Carter 1977, 38) he is leaving behind:

> I would go to the desert, to the waste heart of that vast country, the desert on which they turned their backs for fear it would remind them of their emptiness—the desert, the arid zone, there to find, chimera of chimeras, there, in the ocean of sand, among the bleached rocks of the untenanted part of the world … that most elusive of all chimeras, myself.

However, there are several important problems with the "therapy" of wilderness, as explained by Buell two decades after the "Pastoral Ideology": first, as he insightfully points out (2005, 67), wilderness should be regarded in relative, not absolute terms, because what is an empty, seemingly "untenanted part of the world" (as Evelyn puts it) can be somebody else's "place," a territory of value. Second (ibid., 109–10), there is a tension between this "androcentric construction" of nature as a domain for males as opposed to the "female-coded domestic space" and the traditional coding of nature as a female body, leading to the telltale man/woman = culture/nature formula. Arguing about the paradox, Buell (ibid.) seems to be siding with Alaimo (quoted in Buell 2005, 110) in her "arresting counter-proposal" to recast the ideologically slippery term "nature" as "undomesticated feminist space," developing a cultural critique in which icons like "mother nature" will be replaced by "non-gendered tropes that emphasize continuity between human and nature, while still respecting nature's difference." Eve/lyn's experiences in the desert, while forming around the trope of wilderness, revise the symbolism of mother nature, extending it to the idea of sexuality as "a unity manifested in different structures" (Carter 1977, 66), thereby emphasizing not Alaimo's "non-gender," but gender continuity, with interpenetrability of genders as "correlatives" (ibid., 148) making Evelyn's gender and identity transformation possible.

In terms of the earth–body connection, a series of personifications are used to sustain the idea: it is an "insane" landscape (ibid., 42) of an earth that is "scalped, flayed" (ibid., 41), a world that "shines and glistens, reeks and swelters until its skin peels, flakes, cracks, and blisters" (ibid., 41). It is a specifically female body, which is tired, exhausted: "an abode of enforced sterility, a dehydrated sea of infertility, a post-menopausal part of the earth" (ibid., 40)—in ecofeminist[3] terms, exhausted of resources and suffering: we learn that "somewhere in the vastness" nuclear tests had been performed, spawning "mutations of being" (ibid., 77). Countering "wilderness," the desert is not "empty space" (in Buell's terms)—but rather has lizards rustling in the sand and buzzards floating in the sky (ibid., 42), making Evelyn feel like a frightened foreigner. The desert is also multi-layered: beneath the endless sand and scorching sun, there is a secret underground town, Beulah, built by a matriarch/scientist called Mother, which is both a futuristic and prehistoric realm where high technology combines with goddess-worshipping and ideological proclamation of radical feminism. That the inhabitants—all female—have retreated underground can serve a political cause—away from patriarchy and engaged in guerrilla warfare against it, eventually engaging in an open ground war of everyone against everyone at the end of the book. The name also has mythological and symbolic connotations—Beulah was a poetic name for the promised land of Israel, and is often used in literature to signify an in-between space between heaven and earth—a kind of purgatory, or Middle Earth, where the protagonist is in-between *his* former and *her* future bodily condition. At the same time, Beulah can read as a fallout shelter, an entirely self-sustaining system with synthetic food and recycling of waste, after the ground has become unlivable, and Mother, in her multi-breasted enormity, as a "mutated being," the result of radiation, or a grotesque parody of radical feminism, "the Great Parricide" (Carter 1977, 67) who suffers a nervous breakdown in the end (ibid., 174). She is also a symbol of nature with which Eve/lyn is to unite, when he is led to her, as a captured man in Chapter 6, to be changed into a woman, or as a visiting daughter (the result of the surgery Mother performs underground) in the closing pages: "she was a piece of pure nature, she was earth, she was fructification" (ibid., 60). In what is described as "the cave within the cave" in the technological Beulah (ibid.), she is in a real cave, "the most hermetic of fall-out shelters" (ibid., 180), hiding there "as the witches did in the early Middle Ages" (ibid., 181). Mother's being simultaneously a goddess and a witch, science and nature, black, female, and powerful, avenger and half-mad helpless dying alcoholic (ibid., 189–90) pertains to the fluid, nonbinary conditions she symbolizes. She is but one of the in-betweennesses and transformations that abound in the book, most notably of course Evelyn's own transformation into Eve.

(Trans)gender discourse in *Eve*

As defined by Butler (2004, 6), "Transgender refers to those persons who cross-identify or who live as another gender, but who may or may not have undergone hormonal treatments or sex reassignment operations." Here, "*an*other" is an important improvement on definitions used and understood by the public and even transsexuals themselves. As pointed out by Stone (1992, 222), a transsexual identifies with the " 'opposite' gender" and part of their problem, the sense of being in the "wrong" body, is the tendency to adhere to the binary, without recognizing the distinction of the physical sex and the social gender: "transsexuals commonly blur the distinction by confusing the performative character of gender with the physical 'fact' of sex." A call for a nonbinary "another" is part of twenty-first-century thought, practiced by individuals while still insistently diagnosed as a "disorder." Carter's novel was written in the epicenter of the debates shortly after research into gender dysphoria began in 1968 and resulted in attribution to this state of the official status of a "disorder" in 1980 (Stone 2006, 223). This was soon to be contested by Butler's illumination of gender as performance in 1990, followed by her later discourse on the "ec-static" (2004, 33), as being outside stasis, "beside oneself" (or beside a stable categorizable self) and thus in the realm of the possible, thus refuting the very notion of the "opposite" and thus the "wrong" or "disordered."

Eve depicts two characters who are transgender by definition: Evelyn, who undergoes a male-to-female surgery, and Tristessa, who, though biologically male, cross-identifies as a woman and as a man again later on. Besides, there are several characters balancing on the verge of transidentity—Mother, the matriarch and deity of Beulah, who reinforces her femininity by multiple breast transplants, but whose role as a castrating surgeon is rather emasculate, and who at the end of the novel (Carter 1977, 190) is "old enough to have been either a man or a woman"; the female residents of Beulah, who undergo a ritualistic single mastectomy and whose behavior and lifestyle are relatively gender-neutral; and Leilah who, subjected to hysterectomy and later mastectomy, is one of the most elusive and ambiguous characters we encounter, in both mythological and post-human terms.

The eminence of gender confusion, transgression, and crossing is suggested at the very outset: Evelyn has a premonition of events to come in the queer cinema-goers' admiration of Tristessa for her ability to communicate every nuance of their passions and feelings. Later, he contemplates a seventeenth-century print of "a hermaphrodite carrying a golden egg … the dual form with its breasts and its cock, its calm, comprehensive face" (Carter 1977, 13) in the flat occupied by a Czech alchemist a floor above him in New York. It is interesting that the hermaphrodite was associated with gold, which the alchemist made by mixing different substances. He believed that "the age of reason is over" (ibid., 13) and relished the idea of chaos, "which embraces all opposing forms in a state of undifferentiated dissolution" (ibid., 14). For Evelyn, the symbol of that chaos, of apocalypse, is Tristessa, who he sees as "Our Lady of Dissolution" (ibid., 15). There is something alchemical, anti-rational, and disorderly in the idea of Tristessa *and* the city. With reference to Stryker and Wagner, Straube (Chapter 3, this volume) discusses a complicated relationship dynamics between the transbody and pollution, resulting in the ideas of excess and monstrosity. This is understandable in terms of what Plumwood (2002) describes as the rationalist paradigm, developed from the Cartesian *Cogito ergo sum* "to the extent that rationality is taken to be the exclusive, identifying feature of the human … and that the rational is identified with what is worthwhile" (2002, 98), which leads to human-centeredness and other centrisms (anthropo-, hetero-, Eurocentrism, etc.). Consequently, *any body* that does not fit into the rationalist paradigm is marginalized as excess, waste, and chaos threatening order. Yet, in *Eve*, there is an obvious potential in that chaos, in that blending of the incompatible, of opposites: Evelyn refers to the surrounding chaos as "the entropic order of disorder" (Carter 1977, 15). Entropy implies order coming after chaos, as a result of recombination of what disintegrated, collapsed into a seeming non-existence. Is it that previous assumptions, among other things, were also disintegrating? The book was written in post-Algerian, post-Vietnam, post-Paris'68, post-sexual revolution, postfeminist era of the late 1970s, when many assumptions, especially those concerning gender, were disintegrating. That historical and cultural period, known as postmodernism, links to the present by Baumann's alternative term: "liquid modernity." Opposed to the "solid" nature of earlier areas which were characterized by rigid power structures and value systems, as well as fixed spaces as clear-cut roles, Baumann presents "liquid modernity" (2000, 14), a condition of "lightness and fluidity" of human engagements, unfixed in space and instantaneous in time, "the increasingly mobile, slippery, shifty, evasive and fugitive power" and "the falling apart, friability, brittleness, transience, until-further-noticeness of human bonds and networks." The uniqueness of that fluid society is that it is itself (ibid., 212) "a chaos seeking a form, but a form that is never fixed once for all." For Baumann (ibid., 213), the prerequisite for an autonomous society "and the freedom of its members" is not "fighting the endemic contingency and uncertainty of human condition, but … recognizing it and facing its consequences point-blank." Being in the vanguard of thought and imagination, Carter specializes in this very "contingency," creating tirelessly in-between, fluid, transitory, and, in today's terms, nonbinary, non-rationalist characters and situations. Surfacing out of chaos, they refute the idea of fixedness and "solidity" of identity, body, and self-perception.

Leilah: in-betweenness and natureculture

The idea that one's body does not necessarily incorporate one's identity, that everything is deceptive and masquerade-like, is presented through the character of Leilah, the 17-year-old patois-speaking prostitute Evelyn meets in disintegrating New York. Leilah is a multifaceted character in whom deceptive complexities accumulate: those of gender, humanity/animality, mythology, and corporeality.

Leilah is described as "an in-between thing," between man and animal. Evelyn might be alluding to her blackness—for her blackness and patois dialect are stressed consistently—but also to the idea of a blurring borderline between man and animal. Animal imagery is used in abundance: wrapped entirely in a fox-fur coat, she gave the impression of "a fully furred creature … a witching fox in a dark wood" (Carter 1977, 20); she is addressed as "prey" (ibid., 20, 25); exuding "animal perfume" and "musk" (ibid., 20); singing a high-pitched song like a bird, and seemingly under the influence of an alternate gravity as if flying (ibid., 22) or hovering (ibid., 21); the slums of New York which she occupies are referred to as "habitat" (ibid.). What animal exactly Leilah could resemble puzzles Evelyn, for there is nothing definite about her: "a strange, bird-like creature, plumed with furs, not a flying thing, nor a running thing, nor a creeping thing, not flesh nor fowl, some in-between thing, hovering high above the ground which was, all the same, its reluctant habitat" (ibid.).

In contrast to this or, rather, complementing the image of Leilah, is the insistence on her artificiality, matching that of the megapolis: artificial cream (ibid., 23), hash candy, instant coffee (ibid., 26), nylon, and technicolor: "Her dresses were rags of chiffon or of slimy, synthetic fabrics of harsh-textured knitted, metallic stuff. … Her stockings were made of black, or purple, or scarlet mesh; her vertiginous shoes combinations of shiny leathers dyed green, pink, purple, or orange. She walked in technicolor" (ibid., 29) among the oozing urban rubble "with the rapt delight of a shepherdess in a pastoral straying among flowers in a meadow" (ibid., 21). This oxymoronic description collapses the border between culture and nature, allowing Leilah to spill easily from one into the other.

The next corpus of imagery implies the mythological complexities of Leilah's character, and presents it as enigmatic, dangerous, and negative. Upon viewing Leilah, Evelyn sees her as a "fox pretending to be a siren," a "witching fox," a mermaid (ibid., 22), and a succubus (ibid., 27); when being seduced by her (in fact, himself in lusty pursuit), he feels "the ghastly attraction of the fall" (ibid., 25), in the biblical sense; finally, at the end of his journey and the book, Evelyn (now Eve) learns that the black girl's real name is Lilith, hence the connotations.[4] Finally, we have the idea that Leilah is not only "between" human and animal, the real and the mythological, but that her very flesh does not necessarily contain or reflect all there is to her personality (ibid., 27): "Duplicity gleamed in her eyes and her self seemed to come and go in her body, fretful, wilful, she was a visitor in her own flesh":

> she rolled down the mesh stocking down one black, matte thigh, upon which the coarse mesh had left indentations as tragic as if the flesh had been pressed against barbed wire in an attempt at an escape from a prison camp in which she had always lived, would always want to flee, would always fail.
>
> (ibid., 24)

As mentioned above, much recent critical attention to Carter has been due to its "Butlerification," the increasing attention to the constructed nature of gender in her discourse. Yet we can observe another interesting tendency in this novel: it is not gender that is constructed and, as later Butler (1993) insists, also sex, but the very idea of an identity, a personality locked in the body seems to be undermined. We have, on the one hand, the body as metaphor (with the mythological complexities in this novel) and, on the other hand, the body as never fixed, final, defined, human, or animal, but instead interpenetrable, transgressive, and transcorporeal. Interestingly, it is Leilah who takes Eve/lyn's eventual sex change with "unequivocal acceptance" (Carter 1977, 171), as if their "changed state" (ibid.), Evelyn into Eve and Leilah into Lilith the storm-trooper, is part of perpetual transformations that are but the natural order of things.

Tristessa: (in)authenticity and performance

If Leilah balances on the verge of animality, Tristessa is associated with artificiality from the start: the old movie Evelyn watches preserves her beauty (Carter 1977, 5) "on celluloid" and the essence of her charm has nothing to do with anything "as commonplace as humanity" (ibid., 7). Describing Tristessa as a "fleshy synthesis of the dream" (ibid., 9), he implies his romantic and sadomasochistic fantasies associated with her movies, such as *Wuthering Heights* or *The Fall of the House of Usher*, in which the woman heroine is perpetually suffering and dying. Even her name, coming from the French "la tristesse," sorrow, is emblematic of the "female occupation" she embodies. The idea of embodiment as the actor's job summarizes the discrepancy between the material body, embodiment, and idea (or feeling) in Tristessa's case (ibid., 7–8): "She had been the dream itself made flesh though the flesh I knew her in was not flesh itself but only a moving picture of that flesh, real but not substantial."

The whole array of theatrical imagery employed in depicting Tristessa—her poses, photographs, glass mausoleum with huge tear-shaped sculptures and wax figures of famous Hollywood dead—suggests performativity, theatricality, and "inauthenticity" of her "femaleness" addressed by earlier critics (Makinen 1997; Johnson 1997), posing the "transvestite" as an "incomplete transsexual" (Caroll 2011, 248). Indeed, in Butler's (1990, 175) framework, drag exposes the "contingency of gender," mocking "the distinction between inner and outer psychic space and effectively … both the expressive model of gender and the notion of a true gender identity" (ibid., 174). The advent of transgender theory challenged the notion of "performativity" of the transsexual body (and of gender itself, as in later Butler). As Caroll (2011, 249) rightly stresses, "there is a tension between an insistence that the femininity assumed by the transsexual is culturally constructed, and hence inauthentic, and a presumption that 'women' have a prerogative to femininity based on 'real' female experience." Recent transgender theorists (Prosser 1998; Salamon 2010) move the materiality of the transgender body to the foreground, but in different ways: for Prosser (1998, 65), "embodiment forms an essential base to subjectivity" with skin being "the key interface between self and other," providing the "anaclitic support for the psychic apparatus" (ibid.) and being in the "wrong" body feels like living without skin, damaged and defenseless (ibid., 73). Salamon (2010, 6) disagrees with Prosser on the point that, material as it is, the sense of "what a body is" and "how it is assumed" are collapsed into one self-evident thing. Taking the case of Jan Morris' transition as her example, Salamon (2010, 180–81) analyzes the episode from *Conundrum* when Morris is going through airport security in a "precarious condition" of pre-operative "passing," meaning living and looking like a woman, but still biologically male. Morris' anxiety at what line she will be ordered to by border guards concerns neither the materiality of (his) body, nor (her, liminal) feeling about (her) identity, but "how she is read."

Likewise, Tristessa "reads" on verge of liminality, "felt sense" and "performance," which is not an unusual presentation of drag in culture: analyzing the place of *travesti* in Brazilian culture, Sjöberg (2012, 346) points to its roots in the old carnival tradition, where dressing up as women was part of the spectacle, to "confuse, excite or dazzle the audience" (ibid., 345), though illegal up until the 1940s and a taboo in public life up until the 1960s. He emphasizes the traditionally tragicomic idea of *travesti*, with "a misogynist streak" in the early 1920s, and overemphasis on "female elegance and beauty to compete with the female show-girls' (ibid., 346)—thus, in public imagination, they are always "less" or "more" than "real" women. Carter's interest stems from her experiences in Kabuki theater in Japan in the early 1970s, where men playing women are more glamorous and perfect than women in real life. This is part of an old English tradition as well—take Shakespearean cross-dressing in Elizabethan theater. As Winterson (2016) wittily notes, man-kissing-man in Juliet's clothes is "the most famous teenage sex scene in world culture." In fact, the binary (real/not real, glamorous/tragicomic) works only so long as we do not ask the question, What is a real woman? Are there solid criteria for that, or are they allowed to be fluid in the present-day world? In this vein, when referring to Tristessa as "an unbegotten woman who made no concessions to humanity," Carter (1977, 129) actually refers to how she is read by that "humanity," represented by Evelyns and Zeros or, as Dyer (1993, 12) put it, "the wider process by which any human society, and individuals within it, make sense of that society through generalities, patternings and typifications." As Bhabha (1992, quoted in Sjöberg 2012, 358) insists, this leads to "phobia with regard to notions of transgender as the other."

Tristessa is an elusive personality, of whom little is known except for some details of her decadent tastes, and that (Carter 1977, 105) "she liked best, on vacation, to go into the desert," retreating there completely upon abandoning Hollywood at the age of 40, taking up sculpture in glass (ibid.). That retreat to the desert, her whereabouts unknown both to the media and monomaniacal Zero pursuing her, relates to what Caroll (2011, 246) identifies as "the conventional aim of the transsexual … to disappear: to become invisible, unremarkable and undetectable." That Tristessa chooses to "disappear" in the desert underscores the multiple functions of the desert in the novel. On the one hand, it is a place of "emptiness" matching the "nothingness" of the characters populating it: besides Tristessa, it is where the sterile Zero roams about like the Fisher King of Waste

Land, seeking a murderous revenge on the "Queen of Dykes" (his perception of Tristessa (Carter 1977, 101)), who he believes had "dried up the desert, made it all sand" (ibid.) and "magicked away his reproductive capacity via the medium of the cinema screen" (ibid., 104). Zero's absurdity is but one example of the speciesism and gendered fundamentalism Gaard (2016, 181) refers to when speaking of the harassment of the LGBT community in the wake of Hurricane Katrina, which hit New Orleans in 2006, days before a gay festival was to take place there. In *Eve*, the apocalypse unfolding is also partially blamed on Tristessa. At the same time, her dwelling in the desert is not part of her "nullity"—on the contrary, the "shifting flesh of the desert" (Carter 1977, 107) with mountains "jutting through" it (like a backbone), "capped with snow" (like skin), "hooded with mist" matches her person as transcorporeality that, as I argue, is strongly present in the novel. Retreating to the desert, Tristessa, for whom (ibid., 110) "this world had never been sufficient," goes "beyond the boundaries of flesh" (ibid.)—in Butler's (2004) terms, "beside herself." It is in the desert that Tristessa and Eve unite, mirrors and Doppelgängers, both stretching beyond themselves, beyond the limits of their bodies and the limits of (human) civilization.

Evelyn/Eve's body: technological, mythological, and material

Evelyn's transition, as emphasized by the earlier critics (e.g., Caroll), is involuntary and violent. He is turned into a woman as "punishment," "Leilah's revenge" (Carter 1977, 172), and Mother's grand plan to (ibid., 77) "rejuvenate the world" by Eve's child—save it from the social catastrophe of patriarchal binary and perhaps the ecological/nuclear catastrophe as well. However, there are also deeper meanings to Eve's transformation: he/she is to learn what it is to live inside an "assumed" body, to be an impersonator with a sense of detachment from oneself, to learn to live in a new skin. Just like James Morris comes to "say good-by to myself" in the mirror before the surgery that turned him into Jan (Morris 1975), when Eve looks in the mirror for the first time after the operation from Evelyn, they respond (Carter 1977, 74): "I saw Eve. I did not see myself," and her reply to Mother's question (ibid., 75), "How do you find yourself, Eve?" is "I do not find myself at all." Ironically, though the purpose of punishment is to make Evelyn experience his own wrongdoings toward women—rape, humiliation, unwanted pregnancy—the first effect is realizing a gender dysphoria, a sense that one's body is not one's choice, and one (ibid., 80) "must climb inside the skin of the girl willy nilly, whether I liked it or not, and learn, somehow, to live there." In this respect, Carter's text is not only a (fictional) post-transsexual biography[5] but a fantasy of how a cisgender would feel if they suddenly had to change sex, climb inside a wrong skin—a feeling that transsexual autobiographies explain. Later, as Eve progresses on her rite of passage into womanhood—living as Zero's eighth slave-wife, and finally reuniting with Tristessa, both in a new guise—she comes to sense her old shape as a performance and the memories of (his) past life as (ibid., 92) "old clothes belonging to someone no longer living."

It may seem as though Evelyn/Eve's transition is a straight line, culminating in acceptance and complacency, a solid sense of self that other-sexed people dream of, as in Morris' (1975) early aspiration, when he sensed he was a girl in a boy's body, as he confided in *Conundrum*: "Perhaps one day, when I grew up, I would be as solid as other people appeared to be." According to Johnson (1997, 179), Evelyn's "chronicle" "underlies the appearance of a 'pure' gendered self." However, in my opinion, there is no linear progression toward finality in *Eve*, nor a final sense of identity, but a fluid, circular movement, bringing her to the ocean, which is in no way final, but vast, borderless, and multidirectional. The ocean imagery at the end of the novel, after the urban apocalypse and the sterility of the desert, matches the idea of the body in present-day environmental, gender, and identity theories, allowing us to address the dilemmas discussed in the body–environment discourse.

Blurring boundaries of gender, humanity, and nature

When Tristessa's biological maleness is uncovered, a new set of puzzling issues surfaces. Suddenly realizing "the source of her enigma, of her shame" (Carter 1977, 128), Eve starts forming labels for him in her mind: "a female man" (ibid.), "an anti-being" (ibid., 129), "who represented the refined essence of all images of love and dream" (ibid.). Eve's reaction betrays her/his own ambiguity:she desires Tristessa while still in her feminine guise, thus as a homosexual (since she is already a woman); or perhaps as the former male for whom the transition has not been "complete." Desiring Tristessa as a man after the "exposure," Eve is aware of her own dual nature: on the one hand, she is a perfect woman, and her desire is straight and "authentic." On the other hand, contrary to what earlier critics have claimed, Eve's femininity is no more authentic than Tristessa's. She is described as "a man-made masterpiece of skin and bone, a technological Eve in person" (Carter 1977, 146), while her lover is too complex to be categorized at all: "He, she–neither will do for you, Tristessa, the fabulous beast, magnificent, immaculate, composed of light" (ibid., 143). Neither of them fits into "solid," "human" categories: "You and I, who inhabited false shapes … appeared to one another doubly masked … and now we were no longer human" (ibid., 136). Here the failure to be categorized as "human" should not be viewed in negative terms, although in "solid" culture it would be viewed just this way. Early Butler (1990) has most famously spoken of "abjects" who fail to appear "properly gendered" and thus cannot occupy the "human" categories of subject/object but constitute "the domain of the dehumanized." In the environmental discourse, however, "human" is but one category, no longer the superior one—I have referred to Murphy's "anothers," and Straube (Chapter 3, this volume) also discusses Alaimo's "movement across bodies" to counter the boundary politics, as well as Morton's "polymorphously perverse" intimacy as the interrelatedness and connection with other life forms. Likewise, in Carter's novel (1977, 128, emphasis added), we can observe an initial pronominal confusion in Eve's mind: "I crept up to *him* and kissed *her* pitiful, bare feet with their fine ankles and high ballerina arches … the proud, solitary heroine … the implicit maleness *it* had never been able to assimilate into *itself*," switching thereafter to mythology and animality in depicting both characters.

The myth of Tiresias is alluded to—"I know who we are. We are Tiresias" (ibid., 146)—perhaps to suggest that as transgenders they are not incomplete, or inauthentic, but endowed with the whole of experience. Tristessa in particular, with his long white hair floating "like that of a seer" (ibid., 145), earlier referred to as "serpentine," reminding us of how Tiresias was changed into a woman—for seeing snakes mating. Conventionally understood as punishment, the story of Tiresias can be read as an encounter with nature, and thus transgresses the boundaries of "solid" human desire. There are at least 1,500 species of bisexual animals, and many types of snails, worms, and fish are hermaphrodite. Making love, Eve and Tristessa make "the great Platonic hermaphrodite … the whole and perfect being to which he, in his absurd and touching heroism had, in his own single self, aspired" (ibid., 148).

The reference to "masculine and feminine" as "correlatives which involve one another" (ibid., 149) does not necessarily imply a standard heterosexual couple as it might seem at first sight, but rather that there are masculine and feminine features in all individuals, complete only in nature. Later in this volume, Freyne (Chapter 9, p. 177) speaks of the socially constructed masculine and feminine as two incomplete versions of human nature, while in nature these notions are "in a constant state of fluidity." To understand that, Eve and Tristessa must flee into the desert, leaving the debris of civilization behind. In a condensed apocalypse, Tristessa's desecrated mansion (realm of culture, pretense) perishes in a technological maelstrom, together with her torturers, sinister patriarch Zero, and his obedient self-negating wives.

The scene in the desert is full of magic that depends on the blurring of the boundaries of solid and liquid, human and nature. The stars and the moonlight "melted the gold" (ibid., 150)—of sunshine, but melting of gold has alchemical references (ibid.), the dissolution of solid substances to make a new one. Baumann's ideas of liquidity are helpful once again, only this time (human) nature itself liquefies. When Eve draws furs around Tristessa and herself for warmth, she senses a "unique consolation of the flesh" (ibid.) that has nothing to do with being either man or woman. Carter depicts the inner animal in man in "The Tiger's Bride" (1978, 169), her provocative rewriting of *The Beauty and the Beast*, when it is not Belle who kisses the Beast into humanity, but the Beast who licks "off skin after successive skin, all the skins of a life in the world," leaving behind "a nascent patina of shiny hairs … my beautiful fur," in a union where solid boundaries do not matter. Likewise, Eve senses a bodily merging with the desert: "the breast of a pearl, so white and swollen did the sand look … perhaps we've landed on one of my breasts … as if I too, was the materialization of the moon" (Carter 1977, 151).

Thus, the body is subject to transformation, not only between genders but also different forms of matter. In the desert, Evelyn witnesses the "instantaneous metamorphosis" of a bird shot in the chest—a bird at once reminding one of Icarus or the alchemical bird of Hermes, turning into "dead and putrefying matter" (ibid., 44). There is a clear parallel between the bird and Tristessa, who is also shot in the chest much later: both have respectively feathers and hair "as white as snow," with only the pin-feathers/roots "yellowish, as if tarnished" (ibid., 44, 155); both are buried in the sand, eventually to merge with it, as matter does.

An encounter with the ocean constitutes the final stage on Eve/lyn's picaresque journey, conveying the idea of circularity, non-finality, liquid movement, the "omnivorous inscrutability of the sea" (ibid., 176). At the same time, the ocean is alive and suffering. The apocalypse of New York has now reached the cities of California, the center of civil war and devastation, and turns the ocean into a "refuse bin" (ibid.), a tide carrying "the ignoble detritus of civilisation" (ibid.).

Interestingly, the book ends not only with apocalypse but with a promise of (entropic) rebirth, rendered through an array of natural and naturecultural metaphors. Eve's passing through caves by the sea is both a simulated birth and a compressed version of evolution simultaneous to involution: amber liquefies back into pine tar and pine forests spring upon rocks, while different animal species pass by Eve, who recognizes some of herself in them. At the center is archeopteryx, half-bird half-lizard, a "miraculous, seminal, intermediate being whose nature I grasped in the desert" (Carter 1977, 185)—a reference to both Tristessa and herself, now carrying their child, who "will have two fathers and two mothers" (ibid., 187).

Finally, Eve commits herself "and her little passenger"—the unborn baby—(ibid., 190) to the sea, simultaneously carrying the ocean inside herself—the amniotic fluid, the baby's first environment. Both are an ecosystem within ecosystem, not a final body but part of a cosmic whole.

Conclusion

In this chapter, I have addressed the fluidity and (trans)gender entanglements in *The Passion of New Eve* as related to fluidity of nature, as well as intertwining the body with its environment. The idea of blurring boundaries of culture/nature, of "solid" and "liquid" notions and categories implies blurring of the boundaries of gender and physical sex. The identity dilemmas and gender transformation of the two main characters can be described by such transgender terms as "passing," "transitioning," "reassignment," "dysphoria," and "cross-dressing," but only when viewed within "solid" human categories. If viewed within ecocritical terms, there is no transitioning from one solid form to another, or from one part of the binary to the other, arriving finally at a "true" identity or a final gender, but an endless process of transformation, merging, interdependence on nature and culture, different forms of human and more-than-human. As Parker

(Chapter 1, p. 20, this volume) emphasizes, when trans identities are considered in their relation to the natural world, gender, as well as nature, becomes much more fluid. Similarly, in my chapter, Eve's transformation may be viewed as both a high technological achievement and a sacred ritual act of witchcraft, resulting in a cyborg finding herself through nature, a system of caves, nonbinary love, the change of elements, transformations of time, and eventually the eternity of the ocean, symbolizing the non-finality and liquidity of these forms and notions.

Notes

1 My own shock at my first ASLE/EASLCE conference (Bath, UK, 2010) was the absence of racial diversity.
2 There is still a discussion whether the more accurate term is "transgender" or "transgendered" and what the nuanced differences between these terms are. Butler uses the term "transgendered."
3 Ecofeminists (Warren 2000; Plumwood 2002) link the exploitation of nature by humans to the abuse of the female body, of which rape is the most atrocious manifestation. Land is arid and sterile, because controlled and disciplined like a gendered body. For Plumwood's views, see Parker (Chapter 1, this volume).
4 Adam's mythological first wife, rebellious and fecund, substituted eventually by submissive Eve.
5 As proposed by Johnson (1997, 179), we can read the presentation of Eve's "pre-operative" history as a kind of post-transsexual narrative, in Stone's (2006) terms, as it recounts the preoperative events through the former male's subject position.

References

Alaimo, Stacy. 2010. *Bodily Natures: Science, Environment, and the Material Self.* Bloomington: Indiana University Press.

Baumann, Zygmunt. 2000. *Liquid Modernity*. Cambridge: Polity Press.

"Beulah." *Wikipedia*. https://en.wikipedia.org/wiki/Beulah_(given_name). Accessed July 15, 2016.

Bristow, Joseph and Trev Lynn Broughton (1997) "Introduction." In *The Infernal Desires of Angela Carter*, edited by Joseph Bristow and Trev Lynn Broughton. New York: Longman.

Buell, Laurence. 1989. "American Pastoral Ideology Reappraised." *American Literary History* 1(1): 1–29. www.jstor.org/stable/489969.

Buell, Laurence (2005) *The Future of Environmental Criticism, Environmental Crisis and Literary Imagination.* Oxford: Blackwell.

Butler, Judith. 1990. *Gender Trouble: Feminism and the Subversion of Identity*. London: Routledge.

Butler, Judith. 1993. *Bodies that Matter: On the Discursive Limits of Sex.* London: Routledge.

Butler, Judith. 2004. *Undoing Gender*. Abingdon: Routledge.

Caroll, Rachel. 2011. "'Violent Operations': Revisiting the Transgendered Body in Angela Carter's *The Passion of New Eve*.'" *Women: A Cultural Review* 22(2–3): 241–55. DOI: 10.1080/09574042.2011.587246.

Carter, Angela. 1977. *The Passion of New Eve*. London: Virago.

Carter, Angela. 1978. "The Tiger's Bride." In *Burning Your Boats: The Collected Short Stories*, 154–69. New York: Henry Holt.

Dyer, Richard. 1993. *The Matter of Images: Essays on Representation*. New York: Routledge.

Edwards, Brian. 1998. *Theories of Play and Postmodern Fiction*. London: Routledge.

Gaard, Greta. 2016. "From 'Cli-fi' to Critical Ecofeminism: Narratives of Climate Change and Climate Justice." In *Contemporary Perspectives on Ecofeminism*, edited by Mary Philips and Nick Rumens, 169–92. London: Routledge.

Garrard, Greg. 2004. *Ecocriticism*. London: Routledge.

Johnson, Heather L. 1997. "Unexpected Geometries: Transgressive Symbolism and the Transsexual Subject in Angela Carter's *The Passion of New Eve*." In *The Infernal Desires of Angela Carter*, edited by Joseph Bristow and Trev Lynn Broughton, 166–83. New York: Longman.

Makinen, Merja. 1997. "Sexual and Textual Aggression in *The Sadeian Woman* and *The Passion of New Eve*." In *The Infernal Desires of Angela Carter*, edited by Joseph Bristow and Trev Lynn Broughton, 149–65. New York: Longman.

Morris, Jan. 1975. *Conundrum*. Coronet edition. London: Coronet. www.changelingaspects.com/Life%20Stories/Conundrum.htm. Accessed August 18, 2016.

Murphy, Patrick D. 2013. *Transversal Ecocritical Praxis: Theoretical Arguments, Literary Analysis, and Cultural Critique*. Lanham, MD: Lexington Books.

Peach, Linden. 2009. *Angela Carter* (2nd edition). London: Palgrave.

Plumwood, Val. 2002. *Environmental Culture. The Ecological Crisis of Reason*. London and New York: Routledge.

Prosser, Jay. 1998. *Second Skins. The Body Narratives of Transsexuality*. New York: Columbia University Press.

Salamon, Gayle. 2010. *Assuming a Body: Transgender and Rhetorics of Materiality*. New York: Columbia University Press.

Sjöberg, Johannes. 2012. "Transgendered Saints and Harlots. Reproduction of Popular Brazilian Transgender Stereotypes through Performance on Stage, on Screen, and in Everyday Life." In *The Handbook of Gender, Sex and Media*, edited by Karen Ross, 344–62. Wiley-Blackwell.

Stone, Sandy. 2006. "*The Empire Strikes Back*: A Post-transsexual Manifesto." In *The Transgender Studies Reader*, edited by Susan Stryker and Stephen White, 221–35. New York: Routledge.

Trevenna, Joanne. 2002. "Gender as Performance: Questioning the 'Butlerification' of Angela Carter's Fiction." *Journal of Gender Studies* 11(3): 267–76. DOI: 10.1080/09589230220000021250.

Warren, Karen J. 2000. *Ecofeminist Philosophy: A Western Perspective on What it is and Why it Matters*. Oxford: Rowman and Littlefield.

Winterson, Jeanette. 2016. "Bed Tricks and Broken Women: Shakespeare's Guide to Love." *The Guardian*, April 18. Accessed April 30, 2016. www.theguardian.com/culture/2016/apr/18/bed-tricks-and-broken-women-shakespeares-guide-to-love.

5 Chinese literature, ecofeminism, and transgender studies

Peter I-min Huang

Transgender studies, a relatively new area of critical inquiry, offers useful insights to scholars who specialize in ecocriticism. I make the case for that claim here by way of an ecofeminist reading of several poems in 之間:陳育虹詩選 (*In-between: New and Selected Poems*) (2011), by Taiwanese poet Yu-hong Chen (陳育虹). In addition, I comment on the figure of the goddess Guanyin in the Chinese literary classic 西遊記 (*Journey to the West*) (1592) by Cheng-en Wu (吳承恩). The given figures speak to and for a range of "in-between" conditions, identities, histories, and perspectives that are missing from or erased in mainstream, dominant, and official narratives of Taiwanese and Chinese culture and society. In identifying those conditions, identities, histories, and viewpoints, I draw upon studies by transgender studies' scholars and ecofeminist scholars, since they, more than other critical studies, highlight concerns being voiced in the East about the need for more committed appreciation of moral and affective ties that challenge mainstream reductive dualisms—namely, culture/nature, male/female, and human/animal.

In 現代臺灣文學史,下册 (*A History of Modern Taiwanese Literature*) (2011), Fang-ming Chen (陳芳明), one of Taiwan's most well-known and highly respected literary critics, summarizes the generation of "pioneer" writers to which Yu-hong Chen belongs. They forged modern Taiwanese poetry in the 1980s, when martial law ended and the government opened its doors to so-called free market liberal economic policies and practices. One of the more felicitous outcomes of those seismic political and economic shifts was the "re-revolutionizing" of Taiwanese society and culture by women (ibid., 722). By the latter half of the 1990s, by which time women had more political and economic power under new or revised inheritance and labor laws, women were producing and consuming not only the staple genre of popular romance fiction but also literature that dealt with taboo topics, reflected the influence of postmodern theories of subjectivity, and challenged older and limited chauvinist prescriptions and formations of Taiwanese identity (ibid.). Questioning and rejecting masculinist and putative objective and factual accounts of Taiwanese culture and society, writers and readers called for narratives that explored female sexuality, the role of women in culture and society, domestic life, and affective experience (ibid., 753).

Born in 1952 in the industrial city of Kaohsiung in the south of Taiwan, Yu-hong Chen graduated from the English Department of one of Taiwan's most distinguished institutes, Wenzao Ursuline University of Languages, located in the southern city of Kaohsiung. Thereafter, she traveled abroad and lived in Canada for several years before returning to Taipei. Today, she is recognized for being one of the avant-garde poets who liberated literature in Taiwan from the patriarchal confines imposed upon it during the period of martial law and earlier. Chen was a "late bloomer" and so did not garner critical notice until the 1990s, a decade after many of her contemporaries were being singled out by critics (Fang-ming Chen 2011, 758). However, by that decade, critics were hailing her for being Taiwan's most "lyrical poet," a distinction that reflected Chen's unique use of and focus on sound and rhythm as well as diction (ibid.). In the intervening quarter of a century Chen continued to earn praise, and today she ranks among Taiwan's most critically acclaimed and successful poets (ibid., 758–9).

What is remarkable about Chen's poetry in addition to its exemplary formal lyrical qualities is that it captures or powerfully represents in-between states of identity and erotic modes of being that resonate with both transgender studies theory and ecofeminist theory. Fang-ming Chen, in a critical study entitled 美與殉美 (*Beauty and Sacrifice for Beauty*) (2015), where he devotes an entire chapter to the poet, seems to point toward those connections in his discussion of the titular poem of Chen's collection 之間 (In-between). As he states, the poem represents Chen's literary output as a whole in its confrontation of and resistance to an "outdated mindset of dualism" (ibid., 164). Here are several lines from the titular poem 之間 (In-between):

光影之間，虛實之間，時空之間。
聚散沉浮冷熱動靜去留輕重
以至迷悟死生之間，你我之間。
這介系，這容納我們的方寸。
這幾乎抽象的，因為不確定而極寬大的
一切生發之間。

(2011, 24)

Between light and darkness, between emptiness and reality, between time and space
Between meeting and separation, between surfacing and drowning, between heat and coldness, between activity and stillness, between light and heaviness
Between illusion and disillusion, between life and death, between you and me
This small preposition in-between is a space for both you and me.
Almost abstract, but it has the generosity because of its uncertainty.
Between what happens.[1]

As Fang-ming Chen argues, the poem is fascinating not least because of the audacity with which it opens up gaps between normal continuities and conjoins normative separations. As (Fang-ming) Chen also sums up the critical reception of (Yu-hong) Chen, it reflects that Chen is a poet who is difficult to pin down; notwithstanding the obvious influences of feminism and postmodernism, it does not transparently or easily fit with other contemporary poetry that reflects the same influences.

In another poem, "Let It Rain" (Yu-hong Chen 2011, 26–7), there are these astonishing lines:

> Let the moon perspire softly,
> Let crickets suppress their voices like the flipping of a page.
> Let liquid drop upon liquid.
> Let the wall let the forbidden ground; let the concretization
> let the gap move by itself; let the filling
> let a pen lonely let a fallen leaf scribble.
> Let stardust explore, let memory. (26)

As with "In-between, "Let It Rain" pushes against what Fang-ming Chen calls the "outdated mindset of dualism." I see that resistance to refer specifically to the human/animal dualism. Through use of and attention to touch, sight, sound, smell, and hearing, Chen's poem ingenuously catapults readers into the real worlds, typically unnoticed or dismissed, that are shared between humans and other animals. In another poem, "Cat" (Chen 2011, 760) (a poem that Fang-ming Chen also singles out), the lines "When I and a feral cat try to give us opportunities,/I gaze at her in silence/and we can feel the other's breath./She is no longer a feral cat to me" bring to mind ecofeminist explorations of the ethical as well as corporeal connections between humans and other animals. Disappointingly, Fang-ming Chen and others have addressed neither that aspect of Chen's poetry nor issues of transgender that appear in Chen's poetry.

Fang-ming Chen admits that he is both irritated and fascinated by the poem "Let It Rain" (2015, 167). It evokes for him "female murmurs that mean unbearably nothing" (ibid.). As much as it compels him to reread it, it annoys and stumps him. Simultaneously riveted and piqued by the poem, he asks, "What does [Chen] try to say?" (ibid.). He reiterates his sense of bafflement and intrigue when he comments on these lines: "Let detail, detail, detail be erased … what is left is cold scribble,/Smooth, slippery, let it expand, and let tendril/Let the sounds of the lyre fly and let the ripple of the night … let candle light let tremble without holding back."

Chen's analysis of "Let It Rain" is insightful, authoritative, and nuanced. At the same time, and as his professed irritation suggests, Chen's reading is limited insofar as it reflects heteronormative viewpoints. For example, when Chen presumes that the allusions to sexuality in the poem refer to heterosexual identity and being ("love making between a man and a woman") (2015, 167), he misses and forecloses upon, dishearteningly so, the full potential force of the poem, which admits, opens readers to, and articulates sexualities that not only push past limited and tendentious patriarchal understandings of love and affection between humans but speak in addition for affective bonds and moral relationships that cross species lines. The interest in and theorization of those sexualities is central to the work of both transgender studies' scholars and ecofeminists.

Foremost among the first group is Susan Stryker, one of the contributors to this volume. As she persuasively writes in the Foreword, the recognition of "'unnatural' transsexual embodiment," "monstrous potential," and "egalitarian relationship[s] with nonhuman material being[s]" is inherently transecological (Foreword, p. xviii). It is part of the work of "decentering the tangled webs of trans-huManinimality," encourages us to "fall outside" of kinds of thought and action that the dominant culture leads us to believe are morally unshakable and incontrovertible, and fosters "egalitarian" relationships that cross species lines and challenge "anthropic exceptionalism" (ibid., p. xviii).

Elizabeth Parker, another contributor to this volume, draws upon the work of ecofeminist Stacy Alaimo in her (Parker's) critique of the claim, made famous by Judith Butler, that gender is fundamentally performative. As Parker (Chapter 1, p. 25) states, transgendered identity and being is, paradoxically, both "ontologically inescapable" and indeterminate (Stryker, quoted in Parker). Similarly, Katherine Thorsteinson and Hee-Jung Serenity Joo (Chapter 2), two other contributors to this collection, reference the work of ecocritic Timothy Morton in their defense of transgendered identity and being. Describing it as uncertain, relational, and volatile, they illustrate how ecocriticism and transgender studies work together to encourage more openness to non-heteronormative identities.

Catriona Mortimer-Sandilands and Bruce Erickson's edited anthology, *Queer Ecologies: Sex, Nature, Politics, Desire*, was one of the first studies to bring together transgender studies and ecocriticism. In 2010, the year that it was published, there were few published studies like it except for feminist ecocriticism, or no studies that specifically focused on and critically related ecocriticism to the work of Gay, Lesbian, Bi-sexual, and Transgender (GLBT) scholars. The editors cite poststructuralist thinker Michel Foucault's *The History of Sexuality*, which includes a critique of the regulation of sexuality during the period of modernity and the biopolitical obsession with and codification of heteronormative medical definitions of sex and biological identity (Mortimer-Sandilands and Erickson 2010, 7). Unpacking those definitions, Mortimer-Sandilands and Erickson discuss how mainstream orthodox discourses of nature and sexuality have been linked in an "evolutionary" narrative that "pits the perverse, the polluted and the degenerate against the fit, the healthy, and the natural" and castigates "nonreproductive sexualities" for being "deviant" (ibid., 3, 7). They thus promote an alternative, queer genealogy, one that encompasses an "ecological politics" as well as a "queer politics" (ibid., 2). In doing so, they identify and introduce ecological narratives that question heteronormative definitions, perceptions, regulations, and accounts of both humans and environment (ibid., 5).

The poem "Just for the Plum and Pear Flower Banquet: White Snake" (Chen 2011, 66–75), inspired by the famous Chinese myth of "White Snake," brings to mind the arguments of transgender studies' scholars as well as such ecocriticism scholars as the ecofeminist Stacy Alaimo, the ecocritic Timothy Morton, and the Gay, Lesbian, Bisexual, Transgender, and Queer Studies (GLBTQ) ecocriticism scholars Mortimer-Sandilands and Erickson. It evokes what Mortimer-Sandilands and Erickson call "hetero-ecologies" and "non-normative sexual and gender" perspectives and positions (2010, 22). It also toys with, upsets, and displaces what Mortimer-Sandilands and Erickson characterize as "conventional notions of deviancy and abnormality" (ibid.). It resonates with what Parker (Chapter 1, p. 37, this volume) calls the "trans resistance" to "allegations of inauthenticity" and what Thorsteinson and Joo (Chapter 2, p. 47, this volume) call attention to in their discussion of trans-misogyny. We see those proclivities in the poem, in such lines as: "Deviating from the soul,/I intend to prove the gene of the desire./I want to deconstruct the evolutionary theory,/Uproot the lifeless and rigidized, Dismantle the monument (66). Chen's language challenges trenchant masculinist "survival of the fittest" understandings of evolutionary theory, which privilege heteronormativity, hierarchy, linearity, and competitive and territorial relationships.

"Just for the Plum and Pear Flower Banquet: White Snake" especially brings to mind ecofeminists' critical challenges to institutionalized forms of heteronormative thinking. It speaks for ecofeminism as a whole in the ways it resists patriarchal dismissals of affect (inclusive of emotion), dismissals that say in effect that affect is immaterial because it is not quantifiable. In particular, the poem articulates the argument that Greta Gaard (p. xxiii) makes in the Preface to this collection, as she quotes Mortimer-Sandilands and Erickson: "resistance to heteronormativity" must reflect an "environmental ethics and politics." As part of that argument, Gaard discusses the affinities between transgender studies and ecofeminism under the term "transecology," which refers to practices and ethical positions that are simultaneously "ecocritical, queer, feminist, critically reconstructive and trans*disciplinary" (ibid., p. xx). A transecological ethics thus exceeds "the knowledge boundaries of gender dualisms," explores "the liminal space that might now be called *genderqueer*," and reflects a genealogy, the roots of which are found in the work of ecofeminist theorists, scholars who were among the first to "[queer] the boundaries of gender, species, race, and environments" (ibid.).

"Just for the Plum and Pear Flower Banquet" thus also troubles what Eva Hayward and Jami Weinstein (quoted by Gaard, p. xx) call "ontologized states." It upsets simple and reductive constructed divisions between humans and the natural environments around and constituting "the human." Those divisions reflect profoundly speciesist and anthropocentric attitudes toward the natural world. They are challenged in that poem, a redactive reading of the myth of White Snake. Commonly disseminated versions of the myth represent that White Snake is aberrant, a hybrid creature, an unnatural coupling of a human female and a snake, and a monster that deserves to be punished. Chen refers to those versions in the poem, comparing mainstream narratives that demonize White Snake with versions of similar figures that appear in Western lore. She alludes to the collusion between Eve and Satan, disguised as a snake, in mainstream Christian accounts of the fall of humans from grace; Lilith, one of Adam's partners in apocryphal Christian accounts of that fall who is described as "Satan's other half" and deviant because sexual and so is punished by being transformed into a snake; and Lamia of Greek myth, one of the many women whom Zeus rapes and who is punished for Zeus' crime by being turned into a snake. In Chen's version of the myth of White Snake, White Snake is persecuted and incarcerated precisely because she represents powerful female role models who defy traditional male authorities, have great powers of healing, and work with, not against, nature. In "Just for the Plum and Pear Flower Banquet," White Snake says:

> Don't be shocked;
> don't be confused by my distorted identities.
> I don't lie,
> nor do I have a poisonous fang.
> I do not seek you to devour your blood or eat your meat.
> I came to fulfill my dream.
> Those cold blooded names
> are not me.
> I came
> to shed a tear. (69)

Chen draws readers' attention to the ideological ties between common speciesist representations of animals and stereotypical sexist representations of women. Appearing in literature from both the East and the West, such representations are part of the history of the dismissal of the suffering and oppression of both women and nonhuman animals under anthropocentric, patriarchal, hierarchical thinking.

In Alaimo's "Eluding Capture: The Science, Culture and Pleasure of 'Queer' Animals" (2010), one of the essays in Mortimer-Sandilands and Erickson's anthology, Alaimo emphasizes the "biological exuberance of nature" (31). She emphasizes that, just as it is found in human populations, sexual diversity abounds in human populations, and is a natural not deviant phenomenon. That kind of diversity has been almost entirely overlooked under popular, mainstream understandings of both nature and the human. Animals are divided into three reductive categories: male, female, and the genderless or asexual "it" (which is just one more than the two categories into which humans are generally dumped). Alaimo also draws attention to Haraway's conceptualization of the cyborg, a figure and trope that invites consideration of animals as "'cultural' beings," as well as to another term that Haraway uses, "natureculture," in her (Alaimo's) critique of mainstream, reductive, dualist, and hierarchical understandings of the human species, other species, and ecogenic or natural identity (56, 60).[2]

Many figures in Chen's poetry remind one of ecofeminist defenses of the cyborg and ecofeminist interrogations of the complex ties between what is commonly perceived to belong to nature and what is commonly identified with culture. According to those defenses, beings and things are always somewhere in between both conventional notions of "human" and "nonhuman" and normative definitions of "culture" and "nature." They represent what Alaimo calls nature's overwhelming "sheer inventiveness" and humans' own "queer-green ethics, politics, practices, and places" ("Eluding Capture: The Science, Culture and Pleasure of 'Queer' Animals," 2010, 64). They subtly and imaginatively resist oppressive and heteronormative institutions that suppress pleasure, eroticism, and non-heteronormative sexualities in both the environment ("nature") and "the human." In Chen's description (in the poem "Just for the Plum and Pear Flower Banquet") of Fa Hai, the envious monk who imprisons White Snake (in the Leifeng Pagoda, built in 975, located on the West Lake of Hangzhou Province), one feels the speaker in the poem taking on the full weight of speciesist and patriarchal beliefs, which demonize both animals and women and culturally and materially incarcerate them:

> Stone pillars, obelisk monument, pagoda.
> The same
> All the same
> I crawl on the ground,
> An inch to retreat
> An inch to advance
> My stomach moves against pebbles and twigs
> Moving my scales body
> Meanders, moves straight, tumbles, zigzags
> How can I tell you my position?
> Separate by the wall of eternity
> Piles upon stones of rigid patterns
> How can I let you hear the sounds of cold spring and hot sulfur?
> How can I let you hear the sounds of my soul? (69–70)

Ecofeminist Noël Sturgeon analyzes the complex ideological ties between heteronormativity and the oppression of women and animals in an essay entitled "Penguin Family Values: The Nature of Planetary Environmental Reproductive Justice" (2010). Sturgeon argues that governments must be more critical of industrialized ecologies, especially the "reproduction of industrialized economic systems" and "the high-consuming, decentralized formation of the U.S. nuclear family" (ibid., 118). Those same governments subtly or openly put the blame for problems of "resource scarcity" and "environmental degradation" on "Global South" immigrant populations (who migrate to industrial countries in the North such as Canada, the United States, and the United Kingdom) and the reproduction of those populations (the "reproduction of the 'Others'") (ibid., 125). That blame and cynosure betray heteronormative understandings of "natural" and "normal" based on the suburban, industrial, Christian nuclear family; so, "the vastly greater amount of resources consumed, the reliance of the economy on the exploitation of Global South resources and labor, and the political domination of other countries by the Global North" is "made invisible" by heteronormativity (ibid.).

Giovanna Di Chiro, in "Polluted Politics? Confronting Toxic Discourse, Sex Panic, and Eco-Normativity" (2010), similarly engages with oppressive agendas of heteronormativity in the specific context of the serious issues of pollution and toxic environments. The dominant anti-toxic discourse regularly and typically appeals to patriarchal anxiety about and fear of "chemical castration," "feminization of nature," "pan-species *instability* of maleness," and loss of "*natural* masculinity" (ibid., 201). Certainly, as Di Chiro notes, there are proven links between environmental degradation and infertility and toxic environments and children born with disabilities, and those links must continue to be confronted, but the "sex panic" associated with the awareness of those links egregiously reinforces "heterosexist, queerphobic, and eugenics arguments" (ibid., 202).

In illustration of her point, Di Chiro discusses the work of Janisse Ray, award-winning author of *Ecology of a Cracker Childhood*, about Ray's experience of growing up in a rural, poor, white, Southern community. She discusses in particular a non-fictional work by Ray entitled "Changing Sex," in *Courage for the Earth* (2007), a collection of essays published in honor of Rachel Carson on the centennial of Carson's birth (2010, 219). As Di Chiro summarizes "Changing Sex," Ray follows the writing models of the environmental writers Rachel Carson and Sandra Steingraber and tells of her own encounter with a young couple on a farm in Vermont (ibid., 220). Tracy is a blonde, ordinary woman in her twenties. C.B. is her husband. Ray is confused about the gender of C.B. because "he looked like a woman" and had "a feminine figure and delicate features" (Ray, quoted in Di Chiro 2010, 220). Ray musters courage to ask Tracy and C.B. a "politically incorrect question" about whether they think C.B.'s transgender identity is related to chemical pollution (ibid.). They answer yes and tell Ray of the known links between "endocrine disruptors and the increased numbers of trans people" (ibid.).

Di Chiro praises the kind of work that Ray writes, since it plays a vital role in raising awareness of toxic environments. However, she also notes that that work in many instances "instigates[s] a selective sex panic" (2010, 221). It raises "the specter of environmental contamination causing queer or transgender fears" and so participates in "the reinforcement of compulsory eco(hetero) normativity" and potentially limits "the possibilities for diverse environmental coalitions" (ibid.).

Di Chiro's references to Rachel Carson, the author of the now classic text of environmental literature, *Silent Spring* (1962), especially brings to mind its sixth chapter, "Earth's Green Mantle." In it, Carson describes the "booming 'weed killer' business" and its "plant-killing chemicals" as an industry that domesticates or eradicates what does not obviously or directly serve the human (ibid., 69). She gives the example of a government-sanctioned project of ecocide in the western region of the United States that was focused on destroying sage and other native plants and substituting introduced grasses for those plants. She calls it "one of the most tragic examples of our unthinking bludgeoning of the landscape" (ibid.) "It is no accident," she writes, "but rather the result of long ages of experimentation by nature," that the great plains in North America nurtured and were nurtured in turn by, among many other species, sage, pronghorn antelope, and sage grouse (ibid., 70). The sage plants gave shelter to the grouse for building nests as well as "loafing and roosting areas" (ibid.). In turn, the "spectacular courtship displays of the cocks" helped loosen the ground and aid the growth of the sage and other grasses (ibid.). The sage also provided food for the antelope, especially during the winter, since it is an evergreen plant (ibid., 71). Under eco-normative notions of progress and to "satisfy the insatiable demands of the cattlemen for more grazing land" (or "grass without sage") "millions of acres of sagebrush lands" were destroyed by chemical herbicides (ibid.).

Carson's "Earth's Green Mantle" is useful for reading alongside a poem by Chen entitled "Cursive Script" (2011, 220–22), which is about the renowned Taiwanese modern dance company Cloud Gate and one of its productions, in which dancers "perform" a style of Chinese calligraphy known as "grass" or "cursive" script. Just as the performance connects the movement and character of grass to "grass" script and the human figure, so Chen imagines both sets of figures as equally cultural and equally natural, or as "natureculture" figures:

> On a horizon,
> stretching, turning, twisting,
> a body
> like grass, absorbed fully with the ink of time;
> a meandering awake,
> a dream so wild and savage
>
> A grass which subject to the taming of the wind
> or standing proudly on the edge of a cliff.
> Wind grasped in the palm of the hand, motionless,
> or wind gracefully dancing in a valley-
> A sheaf of grass is revealed by the stroke of a pen
> ...
> a body is like
> a grass.
> stretching forever and ever,
> further,
>
> Still alive (220–22)

The subject of ancient Chinese calligraphy brings me to my conclusion and the subject of a text that predates Chen's poetry by several centuries: the Ming dynasty classic, *Journey to the West* by Cheng-en Wu (ca. 1500–82). Published in Nanjing in 1592, the text is one of the four most famous works of literature from the Ming dynasty (Yu 2006, x–xii). Based on the life of a Buddhist monk, Xuanzang, and his pilgrimage to India in search of Buddhist canonical scriptures, and comprising 1,700 poems as well as prose, Wu's text weaves together "adventure, fantasy, humor, social and political satire, and serious allegory" (ibid., x).

Wu's contemporaries praised the writer for his skills of versatility but also condescendingly treated *Journey to the West* as being so-called light fiction: full of "good humor, profound nonsense, good-natured satire, and delightful entertainment" (Yu 2006, xiii). Today, scholars hold the epic in great esteem, and regularly and frequently consult and study it for its insights into and reflections of ancient Chinese philosophy, traditions, and cultural practices. Many of those are patriarchal, but alongside them is evidence of alternative ways of living and being in the world that emphasize and embrace transgender identity, trans-species relations, and moral and ethical positions that lie outside of patriarchal, heteronormative, and speciesist frameworks. They are being recognized today in particular by transgender studies' scholars and ecofeminists, within as well as outside of China and Taiwan.[3]

One of the most obvious yet least written-about figures in the context of transgender studies and ecofeminism that appears in *Journey to the West* is Guanyin, the goddess of compassion, a deity who variously appears as a male-, female-, and transgendered god/dess in that work and in other famous sources of Chinese myth. Guanyin is a deity who generously and joyfully moves in and out of, and between, genders.[4] The literary critic Quan-zhi Kao brings attention to that "gender fluidity" in a recent article, "The Bodhisattva Guanyin," published in a mainstream widely circulated Chinese newspaper. As he suggests, indeed, it is that aspect of Guanyin (unacknowledged as it is) that makes the deity one of the East's most beloved and perdurable deities (2016, B3). Kao thus comments on several passages in *Journey to the West*, where Guanyin volunteers to fulfill a request by Buddha to recruit a monk to travel to India to find the true scriptures and enlighten the public, and accepts without hesitation a transgendered identity (2016, B3).

What strikes me in addition to Guanyin's gender fluidity in *Journey to the West* is the human–nonhuman animal boundaries that Guanyin constantly crosses and how that challenges demonizations of nonhuman animals in the time of Wu as well as in our current era. Kao emphasizes that aspect of Guanyin when he refers to the episode where Guanyin chides Sun Wukong (Monkey King) for being fearful of monsters and tells him that the difference between bodhisattvas and monsters is a false or nonessential one (2016, B3). The true difference is what Thorsteinson and Joo refer to as an "eco-ethics" (Chapter 2, p. 39, this volume) and what they (Chapter 2, p. 39, this volume) refer to when they describe ethics as a "form of care that does not strictly distinguish self from the other." When Guanyin sees the monk Xuanzang, she immediately recognizes that he is the monk who can accomplish Buddha's mission. Later, a bear-monster, Black Bear Spirit, steals Xuanzang's beautifully embroidered cassock, a gift from Guanyin. The Monkey King (another key character in *Journey to the West*), in a successful attempt to retrieve the cassock, suggests to Guanyin that he be changed into a pill and offered to Black Bear Spirit, and so when Black Bear Spirt takes the pill, he (Monkey King) can make trouble in the stomach of Black Bear Spirit and weaken him. Guanyin stands out among Chinese deities for teaching and practicing extraordinary empathy with and compassion for monster-animals like Black Bear Spirt. When Monkey King is inside Black Bear Spirit, creating havoc and evidently bent upon killing Black Bear Sprit, Guanyin stops Monkey King, saves Black Bear Spirit, and ordains him a god.

As Susan Stryker (2006, 7) notes in the introduction to *The Transgender Studies Reader*, the emergence of transgender studies runs in many instances parallel with the rise of queer studies, and that relationship has been both a close and troubled one.[5] Thus, many transgender theory scholars are now reworking theories of "queer," since they see in those theories "ethnocentrism" (Roen 2006, 656), "homonormativity" residues (Stryker 2006, 7), and still limited understanding of "gendered spaces" (Stryker, Currah, and Moore 2006, 13). They aim to open up the discourse of gender even further than queer theory studies have done (Smith 2006, 320). Yu-hong Chen's in-between gendered figures and in-between animal and nonhuman animal figures resonate with their work as well as with the arguments ecofeminists make about patriarchal and speciesist attitudes toward women and animals. In "Just for the Plum and Pear Flower Banquet: White Snake," Chen defends the love between White Snake and the human Xu Xian. Similarly, the Chinese literary classic *Journey to the West* celebrates Guanyin's transecological embrace of transgender and empathy for "the other," where that refers to minority human populations as well as to species other than the human species. Both the ancient text and Chen's contemporary publication point to beliefs that transgender studies' scholars and ecofeminists interrogate. In response to Stryker's contention that "transgender phenomena haunt the entire project of European culture" (2006, 15), one could add that the figure of the myth of White Snake and the figure of Guanyin, one of the most popular and respected deities in the East (and one of the three most popular deities in Taiwan), are figures open to ecofeminism and transgender studies arguments and bringing those arguments together under the framework of transecology.

Notes

1 Unless otherwise noted, all English translations of Chen's poems are mine.

2 See also Latimer and Miele's discussion of the term "natureculture" (2013). As they summarize it, it represents the transgression of the dominant metaphysical tradition that separates nature and culture and stands for the notion that every material being inclusive of human beings is both culturally and naturally affected by other material beings: "no part of being human … is unaffected by its material interaction with other materialities" (ibid., 11, 16).

3 For example, the critically acclaimed film *Alifu: The Princess* (阿莉芙) (2017), directed by Yu-lin Wang (王育麟), follows the lives of several transgendered individuals who include Alifu, an indigenous Taiwanese, as they struggle with the decision either to come out and face great disapprobation and ostracism or to remain invisible and suffer no less under dominant heteronormative practices. In that struggle they also persevere, connecting with and supporting people who do not fit prescriptive heteronormative understandings of gender. Cleo Woelfle-Erskine observes that trans people throughout the world are making that difficult choice: to remain invisible in full knowledge of the fact that if they come out they will face both exclusion and violence or to come out and risk social exclusion and persecution.

4 For scholarly studies of the transgender of Guanyin, see, for example, 觀音菩薩與觀音法門 (*The Bodhisattva Guanyin and Guanyin Dharma*) (1972) by Nan Huaijin (南懷瑾), who notes that Guanyin is worshiped as both a male and female deity and is "free to practice becoming" according to different situations and contexts (ibid.,14).

5 *The Transgender Studies Reader* (2006) is one of the first studies devoted exclusively to the discourse and issues of transgender. Other important studies include Stephen Whittle's *The Transgender Debate* (Reading, MA: South Street Press, 2000); Paisley Currah, Richard M. Juang, and Shannon Price Minter's edited *Transgender Rights* (Minneapolis: University of Minnesota Press, 2006); David Valentine's *Imagining Transgender* (Durham, NC: Duke University Press, 2007); and Gayle Salamon's *Assuming a Body: Transgender and Rhetoric of Materiality* (New York: Columbia University Press, 2010).

References

Alaimo, Stacy. 2010. "Eluding Capture: The Science, Culture, and Pleasure of 'Queer' Animals." In *Queer Ecologies*, edited by Catriona Mortimer-Sandilands and Bruce Erickson, 51–72. Bloomington and Indianapolis: Indiana University Press.

Alifu: The Prince/ss (阿莉芙). 2017. Dir. Yu-lin Wang (王育麟); Dist. Swallow Wings Films.

Carson, Rachel. 1962. *Silent Spring.* London: Penguin Books.

Chen, Fang-ming (陳芳明). 2011. 現代臺灣文學史,下册 (*A History of Modern Taiwanese Literature*, Vol. II). 台北：聯經出版社 (Taipei: Linking Publishing Co.).

Chen, Fang-ming (陳芳明). 2015. 美與殉美 (*Beauty and Sacrifice for Beauty*).台北：聯經出版社 (Taipei: Linking Publishing Co.)

Chen, Yu-hong (陳育虹). 2011. 之間:陳育虹詩選 (*In-between: New and Selected Poems*). 台北: 洪範書店 (Taipei: Hong-fang Books).

Di Chiro, Giovanna. 2010. "Polluted Politics? Confronting Toxic Discourse, Sex Panic, and Eco-Normativity." In *Queer Ecologies*, edited by Catriona Mortimer-Sandilands and Bruce Erickson, 199–230. Bloomington and Indianapolis: Indiana University Press.

Kao, Quan-zhi. 2016. "The Bodhisattva Guanyin." *The United Daily*, February 10.

Latimer, Joanna and Mara Miele. 2013. "Naturecultures? Science, Affect and the Non-human." *Theory, Culture & Society* 30(7–8): 5–31.

Mortimer-Sandilands, Catriona and Bruce Erickson. 2010. *Queer Ecologies: Sex, Nature, Politics, Desire*. Bloomington and Indianapolis: Indiana University Press.

Nan, Huaijin (南懷瑾). 1985. 觀音菩薩與觀音法門 (The Bodhisattva Guanyin and Guanyin Dharma). 台北：老古文化公司 (Taipei: Ancient Culture Co.).

Richard, Timothy, trans. 2008. *The Monkey King's Amazing Adventures.* Rutland, VT: Tuttle Publishing.

Roen, Katrina. 2006. "Transgender Theory and Embodiment: The Risk of Racial Marginalization." In *The Transgender Studies Reader*, edited by Susan Stryker and Stephen Whittle, 656–65. London: Routledge.

Smith, Brice. 2006. Book review of *The Transgender Studies Reader*, edited by Susan Stryker and Stephen Whittle. *Trans-Women Studies Quarterly* 36(3–4): 318–20.

Stryker, Susan. 2006. "(De)Subjugated Knowledges: An Introduction to Transgender Studies." In *The Transgender Studies Reader*, edited by Susan Stryker and Stephen Whittle, 1–17. London: Routledge.

Stryker, Susan, Paisley Currah, and Lisa Jean Moore. 2008. "Introduction: Trans-, trans, or Transgender." *Trans-Women Studies Quarterly* 36(3–4): 11–22.

Sturgeon, Noël. 2010. "Penguin Family Values: The Nature of Planetary Environmental Reproductive Justice." In *Queer Ecologies*, edited by Catriona Mortimer-Sandilands and Bruce Erickson, 102–33. Bloomington and Indianapolis: Indiana University Press.

Wu, Cheng-en (吳承恩). 1592, 1972. 中國古典名著：西遊記 (Chinese Classic: *Journey to the West*). 台北：三民書局 (Taipei: Sanmin Bookstore).

Yu, Anthony C. 2006. *The Monkey & The Monk: An Abridgment of* The Journey to the West, edited and translated by Anthony C. Yu. Chicago, IL: University of Chicago Press.

6 Gendercrossing at the frontier

Annemarie Schwarzenbach's transgender memoirs in the Alborz Mountains

Mat Fournier

To think in terms of transecology is to dwell on a paradox, or more accurately on a series of paradoxes. At the intersection of the fields of transgender studies and environmental studies, transecology aims to use the critical tools of both fields to look at the complex apparatus made up of discourses and living things we call "nature." The most obvious paradox in this endeavor is the fact that we, transgender human individuals, are traditionally perceived as unnatural (Stryker, Foreword, p. xviii, this volume). Discussions of the gender binary are so enmeshed in ideas of the "natural," and descriptions of biological phenomenon are so informed by the ideology of sexual difference,[1] that any human body distancing itself from it falls under the category of the abnormal. As a matter of fact, transgender individuals are caught in a double bind. Without the use of gender-reassignment technologies, hormonal therapies, or surgery, we appear as aberrations of nature: incongruent bodies, ambiguous physical traits, or social attitudes. But when we rely on those technologies, which sometimes allow us to pass as "normally gendered" individuals, we become artificial creatures again, techno-dependent cyborgs estranged from the realm of nature.

The second paradox pertaining to transecology relates to the notion of ecology. As an object of study, ecology relies on a consistent and stable network of (natural) relationships. The world of interactions taking place in any milieu, which ecology, as a science, aims to map and analyze, would lose its significance if it were transient, if it had the unstable quality evoked by the prefix "trans-." While the idea of nature is associated with permanency and self-sufficiency, more often than not in a prescriptive manner, transecology raises the question of the relationship between nature and technology, on the one hand, and of the timeless nature of "nature," on the other. As its rejection of transgender bodies, the ideological construction of a stable and quintessential nature ignores the deep, pervasive, and ever-changing entanglement between the biological and the technological. This is precisely why transecology offers an invaluable tool to describe the relationship between human animals and their environment. As a critical tool, transecology enables us to see beyond two binarisms hindering our understanding of the living world as well as the possibility for gender fluidity: nature versus technology; stability versus mutation.

This chapter uses the concept of transecology to describe the relation between the transgender body and the ecosystem of a high-altitude landscape, as it appears in *Das glückliche Tal* (*The Happy Valley*), a 1939 novel by Swiss writer Annemarie Schwarzenbach. In so doing, I hope to disrupt the binary misconceptions shaping understandings of gender deviance and its relationship to the natural world, as well as to demonstrate the relevance of transecology as a critical tool.

"To make me healthy, they took me to this valley."[2] By these words, the narrator of *Das glückliche Tal* (2006, 26) recalls the reasons for his presence in the Iranian Lar Valley, where, for one summer, he tries to get over the dramas of the past while finding the strength to look toward the future. In the Lar Valley, on the "roof of the world," the narrator will attempt to build the transecology of a livable space, of a possible relationship between body and landscape. This chapter describes how the high-altitude settings constitute an organic part of the narrative, allowing the narrator to explore gender deviance. The Lar Valley constitutes a retreat, literally situated above humane society, in a manner that can be compared to Stryker's (Foreword, this volume) evocation of the Icelandic Highlands. The upper valley's environment is eventually the only place where the narrator can find himself: a trans space, "perpetually precarious" (Stryker, Foreword, p. xvii, this volume), a frontier beyond or above civilization.

Annemarie Schwarzenbach never hid her preference for women and is therefore mostly read as a lesbian, a label she neither appropriated nor denied; but I opt here for a trans reading of her work. In doing so, my aim is to shift the perspective from sexual orientation to gender incongruence, and to examine how this disruption operates in her writing. I do not wish, however, to categorize her as a trans man, if only because I do not believe that this assignation has any relevance when not reclaimed by the subject itself. Moreover, Schwarzenbach lived at a time when some individuals, albeit rarely, underwent sex-reassignment surgery and lived under a gender identity different from the one assigned to them at birth. (See the case of Lili Elbe in Parker, Chapter 1, this volume.) While Schwarzenbach often wore men's clothing, she attempted to pass as male only occasionally (Alexis Schwarzenbach 2007). Therefore, I use female pronouns when referring to her.

Although most of Schwarzenbach's works have not been translated into English, she has achieved, decades after her early death in 1942, a notoriety stemming from her legend much more than from her works. Although she was a talented fiction writer, and a prolific reporter and photographer, she remains known as an iconic and androgynous character doomed by a tragic destiny. As I will argue, what has been called the "Schwarzenbach myth" needs to be understood in relationship with gender deviance. The myth didn't appear after her death but has accompanied her all her life, turning her asserted masculinity into a "mysterious ambiguity." The first part of this chapter goes back to the formation of this myth in order to introduce Schwarzenbach and her works to an American audience, and to unfold the relationship between gender variance and the difficulty of inhabiting a social space.

The second part of this chapter describes how the Orient, apprehended as a frontier, provided Schwarzenbach with a geographical equivalent of social inadequacy. A disorienting space, the Orient was a place where getting lost made sense, where the difficulty of inhabiting a space of one's own became visible. I then turn toward the high-altitude environment described in *Das glückliche Tal*. There, the narrator is facing a never-ending challenge: to survive in an inhospitable environment, "almost uninhabitable," as the Iceland Highlands. It is, however, in this barely livable space that the trans body can be felt, or understood, in an almost livable way, far above the discriminating structures that shape the social world.

In my conclusion, "Frontier is the Heart," I will show how, despite their inhospitable ecology, mountain environments are a place of new beginnings. As much as they are a border, they are also a heart—a center from which the rediscovered trans body can finds its way toward the world surrounding them. Schwarzenbach's transecology draws from an *inhuman* nature to map new human territories.

The Schwarzenbach myth

Even though she achieved some notoriety in her lifetime, mostly as a travel journalist and photographer, the largest part of Schwarzenbach's fictional works were published decades after her death, throughout the 1990s and 2000s. This late literary recognition has been both triggered and overshadowed by the celebrity of the writer and by the aura of drama surrounding her. As Schwarzenbach critic Walter Fähnders points out, the romanticizing of her biography often prevented a grounded analysis of her work:

> [T]he oeuvre is completely overshadowed by her biography, which is admittedly fascinating enough and has provided material for films, plays, and novels. A typical example is the following racy sentence which can be found in *Cosmopolitan* under the heading "Life on the Fast Lane" and that strives to sum up an entire life: "She was highly gifted, rich, restless, passionate, loved Erika Mann, and died at the age of 34." (Anonym 2008)[3]
>
> (Fähnders 2010, 20)

Those dramatic interpretations didn't start after Schwarzenbach's death—in fact, they began in her youth and paralleled her first publications. Born in Zurich, Switzerland, in 1908, Schwarzenbach belonged to a very visible upper-class family of this conservative city. As early as her teenage years, her boyish manners and unapologetic preference for women raised local scandals. The young Annemarie wore her hair short, dressed as a boy, rode the family racehorses, and fell dramatically in love with her female classmates, friends, and cousins. To make things more complicated, Annemarie's own mother, Renée Schwarzenbach, was happily sharing her affection between her husband, the wealthy factory owner Alfred Schwarzenbach, father of her five children, and her mistress, Nazi opera singer Emmy Krüger (Alexis Schwarzenbach 2007, 84, 113).

Renée, a Prussian aristocrat and a proud supporter of Adolf Hitler, didn't feel there were any contradictions between being a dutiful spouse and entertaining a "passionate friendship" for a woman. Renée's long-lasting love affair with Emmy Krüger offers a remarkable example of the glass closet conceptualized by Sedgwick (2008, 67–90). Emmy Krüger spent months at a time in the Schwarzenbach household, sharing Renée's bed, and having her say in the children's education (Alexis Schwarzenbach 2008, 9–60). As can be read in their extensive correspondence, Emmy and Renée also shared conservative ad heteronormative values regarding gender roles and family structure. They felt, in a seemingly blissful agreement with Renée's husband, that privately shared feelings, intimacy, and sex didn't prevent the rule of lawful patriarchy. Renée and Emmy's arrangement never undermined the rule of the Schwarzenbach clan. Family was their core value, and its symbolic order was strangely threatened by Annemarie's teenage love affairs.

At the age of 13, Annemarie was already engaged in publicly claimed and *named* same-sex romances. She was also already at war with her family—or was it the other way around? The quarrels provoked by her teenage affairs marked the beginning of her reputation as an "enfant terrible," and, most significantly, planted the seed of the suspicion, shared in Renee and Emmy's correspondence, that Annemarie was mentally impaired (Alexis Schwarzenbach 2007, 171–2, 253). This seed was to grow into a series of commitments in various institutions, at the family's instigation and sometimes without the patient's consent, throughout Schwarzenbach's life. As I will discuss later, this process eventually played a part in her untimely death at the age of 34.

The association between homosexuality and/or transsexuality (which were at the time perceived as contiguous phenomena) and mental disease was in no way exceptional. Unique to Schwarzenbach's case was the fact that she was institutionalized by her lesbian mother. But what might seem like a paradox needs to be understood through Sedgwick's concept of the glass closet. Renée and Emmy's innocence was threatened by Annemarie's visibility as a lesbian, *and* as a masculine woman. To protect Renée's invisibility, Annemarie had to be made visible—as dysfunctional. Even before she had any interactions with the outside world, she had already been cast by her family as a lost cause, a character which then evolved into that of the "broken angel" described by friends, lovers, readers, and biographers.

In 1931, the 23year-old Annemarie, having earned a doctorate in history, left Zurich for Berlin, where she planned to pursue a career as a writer and to act freely on her attraction to women. She published her first novel, *Freunde um Bernhard*, in 1931, and the following one, *Lyrische Novelle*, in 1933. These publications, along with her friendship with Klaus and Erika Mann, children of Nobel Prize winner Thomas Mann, who was well acquainted with European literary circles, allowed her to achieve a certain notoriety. But like the Mann children, her accomplishments were preceded by an aura of seduction and scandal created by her youth, her unapologetic homosexuality, and her use of drugs.

Schwarzenbach's reputation, and its implicit link to gender disruption, can best be summarized with one world: "angel." This attribute is used over and over to qualify her, in various occasions and in several languages. French writer Roger Martin du Gard dedicated his book *Confidence africaine* to her with the words: "For Annemarie Schwarzenbach, in thanking her for walking on this Earth with the beautiful face of an inconsolable angel"[4] (1931). Thomas Mann describes her in his diary as a "ravaged angel" (Alexis Schwarzenbach 2007, 196). The angel metaphor comes, according to photographer Marianne Breslauer, from Annemarie's androgynous type of beauty. Breslauer, who added to Schwarzenbach's reputation with several series of portraits, recalls: "Annemarie had this effect on me then, that she had on everybody: this curious mixture of man and woman .... For me ... she looked like I had imagined the archangel Gabriel in Paradise"[5] (Bonstein 2000).

This description is key to what Fähnders calls "the Schwarzenbach Myth" (Fähnders 2010, 32). While lending a romantic aura to the character, the image both brings out and undermines Schwarzenbach's gender ambiguity. As Breslauer states, Schwarzenbach's seductiveness is rooted in ambivalence: a mixture of man and woman. Angels, as we know, have no sex; that is, no gender and not much sexuality. The very romanticism of the image erases the potential threat of gender disruption. In other words, Annemarie is an androgynous and mysterious creature, but not a hardcore butch, not a female-assigned human being uncomfortable in her social role, not a lively human with sexual needs or desires. Her disruptive body, which is also sometimes described as too thin, too athletic, too masculine, is blurred into the ethereal dimension of the angel.

Fallen angel: from androgyny to erasure

In her literary works, Schwarzenbach alternatively uses male and female pronouns, or maintains the "neutrality" of her narrator by avoiding pronouns altogether (Rohlf 2010, 167). The novella *Eine Frau zu Sehen*, for instance, evokes the narrator's love for a woman. Since, in this case, the whole plot revolves around lesbianism, the reader understands that the narrator is female-bodied. But she refers to herself by the German noun *Mensch*, "human," which can designate both men and women. At some point the narrator looks at herself in a mirror and describes her reflection as someone she could like "as one loves a younger brother"[6] (Schwarzenbach 2008b). In her journalistic work she uses, as sparsely as possible, female pronouns. When her writings assume the appearance of fiction, they feature a male-identified first-person narrator, such as in *Lyrische Novelle*, her second novel. But most of her autobiographical writings are deprived of traceable gender markers, while *Das glückliche Tal* features a male narrator identified with the writer.

Although most critics and biographers qualify Schwarzenbach as gender "ambiguous" (Rohlf, Decock), I argue that her gender fluidity, far from orienting her toward androgyny, can be read in terms of masculinity. In *Female Masculinities*, Halberstam writes: "For a large part of my life, I have been stigmatized by a masculinity that marked me as ambiguous and illegible" (1998, 19). The same remarks can be made about Schwarzenbach, even though in her case illegibility has been turned into a romanticized androgyny, erasing the masculinity and reassigning the subject as a victim. The obvious misogyny of this process—reading nonconformity into a drama, or a disease—is no different from, and is indeed related to, Annemarie's designation as mentally ill by her family. Masculinity, Halberstam claims, doesn't belong to cisgender men but its appropriation by others is often deemed socially unacceptable (1998). Androgyny, on the other hand, is by definition neutral and unthreatening, particularly when accompanied by youth and drama. For reasons mentioned above, I do not intend to reclaim Schwarzenbach as a transman. My point here is that the image of androgyny assigned to Schwarzenbach was indeed destructive in terms of sexual orientation and gender identity—both for the lesbian and the masculine subject.

In 1935, while Annemarie was undergoing treatment for her morphine addiction in the clinic led by Swiss psychiatrist Oscar Forel, he declared his patient to be suffering from "schizoid personality disorders" (Alexis Schwarzenbach 2008, 215). The diagnosis was confirmed in 1938 by another reputed psychiatrist, Ludwig Binswanger,[7] who wrote in his report that his patient was suffering from "an insidious form of schizophrenia" (Alexis Schwarzenbach 2008, 261). Forel and Binswanger justified their diagnosis with symptoms related to gender dysphoria and homosexuality: "lack of restraint and modesty," and "indecency" (Alexis Schwarzenbach 2008, 216). Those terms mostly referred to their patient's refusal to wear feminine clothing, and to her being sexually and sentimentally attracted to the "wrong" object.[8]

In September 1942, Schwarzenbach was the victim of a bicycle accident. The subsequent head trauma plunged her into a coma. She died two months later, in November 1942, and the accident was considered to be the cause of her death until Annemarie's great-great nephew, the historian Alexis Schwarzenbach, started working on the family archive. He found out that after the accident Annemarie was put back in the care of Forel, who had diagnosed her with schizophrenia several years before. When she awoke from the coma induced by her accident, Forel, confident in his schizophrenia diagnosis, reverted to his treatments of election: insulin shots and electroshock therapy. She died two months later, after what Alexis Schwarzenbach describes as a case of spectacular medical negligence. This abusive treatment is without doubt related to Annemarie's disruptive sexuality and gender identity. A tautological and devastating conclusion, her death was characterized by her mother as "the result of an illness she suffered from all her life" (Alexis Schwarzenbach 2007, 310).

In a 1935 letter to her friend Klaus Mann, Annemarie Schwarzenbach ponders the possibility of a life in Switzerland, answering with a no. "Switzerland is a small country, and, should I remain unmarried here … I would remain the target of the preoccupation, the ill will, the fractional hatred, and the lust for sensation of the 'Society'"[9] (Breslauer and Wanner 1997). Unmarried: that is, out of the heteronormative patriarchal society. If questions of body, space, and environment are so crucial in Schwarzenbach's works, it is because she was recurrently deprived, physically and mentally, of a space of her own. It was only through her written accounts that she could design a breathable space. Schwarzenbach's transecology is, unfortunately, a textual one.

Disoriented orientalism

In the introduction of their edited volume *Inside Out*, critics Sofie Decock and Uta Schaffers (2008, 7) describe Schwarzenbach as a "figure of the in-between." According to them, she escapes every attempt that would place her in a given category. Schwarzenbach, they write, "defies all types of categorizations, in her patterns of travel, her de-stabilization of identity (not only gender identity) as well as in her experimentation with textual spaces between facticity and fiction" (ibid., 8). The resulting ambivalence reveals the "repressed potentiality of the in-between" (ibid.), which makes Schwarzenbach's writing so powerful. Decock and Schaffers quote Kleist's *Penthesilea:* "As far as I know is there in nature only strength and its opposition, and no third way,"[10] says Ulysses while discovering the fighting Amazons (Kleist 1998, 3). According to Decock and Schaffers, Schwarzenbach's in-between position allows her the creative freedom of "a third way"; that is, of a disruptive dynamic that challenges the establish field of "nature" and "strengths."

While I agree with the productive side of the in-between position as described by Decock and Schaffers, I want to argue that this "third way" is less a path than a flight,[11] always threatened by erasure or reassignation. For Schwarzenbach as well as for the Amazon queen, it is a matter of finding a space of one's own, a space that doesn't exist and must be created, or won, repeatedly. Like war for Penthesilea, the twin experience of writing and wandering is of vital importance in Schwarzenbach's life and is about the quest for a living space.

Geographically, Schwarzenbach was constantly on the move. In her early twenties she was already traveling throughout Europe, staying in Berlin, Paris, Venice, or Vienna and coming back regularly to Switzerland. In 1933, fleeing the rise of Nazism, she left Berlin and embarked upon what was to be the first of a cycle of travels through the Middle East. First traveling from Syria to Persia with a group of archeologists, she returned to Tehran two years later, in 1935, and traveled through Iran, camping for a few weeks in the Upper Lar Valley, the setting of *Das glückliche Tal*. These cycles of "Orient" travels—I will come back to this term later—provided the material for a variety of texts. First, in 1933, Schwarzenbach wrote a series of articles for the Swiss newspaper *Zürcher Illustrierte*. In 1934, she published a travel diary under the title *Winter in Vorderasien* ("A Winter in the Middle-East"), then worked in 1935 on a short-story cycle, *Bei diesem Regen* ("In This Rain") and finally produced two novellas, *Tod in Persien* ("Death in Persia," written in 1936) and *Das glückliche Tal*, written in 1938. As Decock (2010, 45) writes, "Schwarzenbach's texts on the Orient can be described in their entirety as a highly intertextual network, in which texts can speak to each other because of their intricate connections with regard to space and characters."[12] The density of this material proves how deeply the idea of Orient is connected to Schwarzenbach's writing method. The experience of otherness and of wandering, indistinguishable in her perspective, triggers the creation of yet another space, this one virtual.

Schwarzenbach's narrative use of a foreign space is framed by orientalism (Euchner 2008, 132). Schwarzenbach fails to perceive the countries, landscapes, and cities she is describing as something other than a frontier, a space defined by its difference from the one to which she is accustomed. Euchner writes: "This failure is also noticeable in her text, in which she taints the space of the absolute Other, as which Persia and its customs present themselves to the traveller, always with the vocabulary of the European—i.e. the colonizer."[13] Schwarzenbach's Orient may be compared to *Heart of Darkness*' Africa, or to André Gide's Mahgreb—Schwarzenbach sometimes compared herself to the narrator of *The Immoralist*. The possibility of loss offered by this foreign space is most threatening, and most rewarding, to those who are already lost to society—meaning, to Western society. "Native" societies are being treated as part of a dangerous, inhospitable landscape. The Orient is the opposite of Switzerland insofar as one can be forgotten, or get lost, even to oneself.

The fever trope offers a telling example of this narrative orientalism. In Schwarzenbach's Orient short stories, as well as *in Death in Persia* and *The Happy Valley*, fever is a recurring theme. As a rule, all characters (most of them being fellow Europeans and Americans) fall sick with fever. In the short story *Bei diesem Regen*, the narrator encounters a military cartographer stationed in desolate hills near Aleppo. Bedridden, the young soldier shares his feelings of utter powerlessness: a constant rain is triggering bouts of malaria; but the fever, he claims, is easier to endure than the surrounding hills he needs to map. "I don't understand anything about hills, he says, anything about this country. Everything is so difficult"[14] (2008a, 43). The very geography of the country (the hills), and the climate (the rains) are triggering the soldier's disease, which will, eventually, kill him. Through the recurring motive of the fever, the Orient is made into a condition: the displaced Western subjects are literally lost, in the world and to the world.

In her *Queer Phenomenology*, Sara Ahmed (2006, 6) questions the notion of disorientation, and its consequences for social life:

> When we experience disorientation, we might notice orientation as something we do not have. … It is in this mode of disorientation that one might begin to wonder: What does it mean to be orientated? How do we begin to know or to feel where we are, or even where we are going, by lining ourselves up with the features of the ground we inhabit, the sky that surrounds us, or the imaginary lines that cut through maps?

The Orient, for the European traveler, is a disorienting space. The Orient is the ground where disorientation begins, since it is the place of Otherness. The loss of familiar features corresponds to the failure described by Euchner. As many of her fellow travelers, Schwarzenbach tends to essentialize her disorientation by inscribing it into the very landscape of the Orient. Orientalized, the Orient becomes another word for disorientation. On a daily basis, Schwarzenbach is unable to inhabit the Orient in the way she would inhabit Europe—*if* she could stay in Europe. Facing the landmark of Mount Damavand's familiar silhouette, the narrator of *Das glückliche Tal* reflects: "Maps lie. They know only *one* aspect, and in the cross of North, South, East, West, Mount Damavand remains one and the same. But I have seen another mount Damavand"[15](63).

Maps are lying: this is the Orient's revelation. "But 'getting lost,'" Ahmed (2006, 7) writes, "still takes us somewhere; and being lost is a way of inhabiting space by registering what is not familiar: being lost can in its turn become a familiar feeling." For Schwarzenbach, the disorienting revelation is the continuity of a previous experience, since her familiar European country didn't provide her with a livable space. In the "home" country, life is not an option, because of the social violence the subject is subjected to—a "target," forever exposed, rendered visible by her inadequacy. In the frontier space of the Orient, everyday life itself becomes a challenge. The climate, the language, the landscape, the food, trigger feelings of strangeness. Isn't the Orient, then, the most natural place to navigate? If, as Ahmed (2006, 7) states, "the work of inhabiting space involves a dynamic action between what is familiar and unfamiliar," then traveling in foreign and "inhospitable" spaces feels strangely familiar for the gender-nonconforming subject. In fact, the uncanny becomes easier to deal with when transferred in the surrounding landscape, instead of being located in the subject itself. Or it seems, at least, in literature. Schwarzenbach's Orient writings are haunted by a displacement: invisible social violence as fever; feelings of inadequacy as disorientation; lack of agency as disorientation. The Orient, as a milieu, allows her to make visible the uncanny.

Interlude: heart of Europe

As Decock (2010, 137) points out, the Middle East has a deep and hidden kinship with Europe, or what Schwarzenbach refers to as "the European heart." Decock traces this kinship to nineteenth-century German orientalism, inspired by Protestant theology and early studies of Indo-European culture. Persia plays a key part in this phantasmatic history, through the figure of Alexander the Great and the city of Persepolis. Schwarzenbach, while traveling with a group of archeologists, spent a few weeks in Persepolis. Her arrival in sight of the city is a moment of solemn wonder, related in several of the Orient texts. "Persepolis was at the end of a new plain, on a high terrace its pillars pointed wonderfully into the cloudy night sky, and the name was made reality"[16] (Schwarzenbach 2008d, 164). Persepolis could as well mark the end of the world, almost out of reach on the other side of yet another plain. But once achieved, this mystical goal of every traveler and archeologist becomes reality—the Orient is their troubled reality.

In Schwarzenbach's writing, the imaginary kinship between Europe and the Middle East always appears along with another recurring motive: endless discussions between the narrators and their travel companions, focusing on Europe. "We remained in the vast garden of an English friend the whole day, thinking of Europe"[17] (2008d, 165). Traveling to or staying in Persia brings to light a kinship among Europeans: not only the fact that they are strangers and colonizers, but above all, after 1933, their shared concerns about Europe's future. Nostalgic, these conversations are also embedded in the grim reality of day-to-day political developments and the looming possibility of a war. The homesickness shared by the travelers is anchored in this grim reality. As the Persia of Alexander, or the biblical cities the archeologists are looking for, the Europe they "think" about is forever gone. They are homesick for a time as well as for a place.

Inhospitable ecologies

As Ahmed underscores, inhabiting space is a form of labor, one we often overlook since we are usually treading on familiar ground. But to anyone who travels to a high-altitude environment, the labor of simply being here becomes perceptible. The rarity of oxygen has immediate effects on the body, and it can take several weeks before acclimatization. The upper Lar Valley camp where Schwarzenbach was staying was elevated in excess of 12,000 feet, which qualifies as a "very high-altitude" zone. Upon their arrival at this elevation, most individuals will experience more or less acute symptoms of mountain sickness, ranging from shortness of breath and rapid pulse to insomnia, dizziness, headaches, and loss of appetite.

The first pages of *Das glückliche Tal* describe the "work" of inhabiting the valley as an exhausting and constant labor. No bodily function can be taken for granted, not even breathing or heartbeats. Camping in the upper valley and absorbed in his contemplation of the surrounding cliffs, the narrator is torn between exaltation and discomfort. These sensations translate as fascination and horror for the landscape he is facing:

> When one stands somewhere in the middle of such a slope … one can hear the incessant rustling noise of the debris quite distinctly. This extremely light, monotonous rustling is the only sound in this desolation …. I know of no more intolerable sound than this unceasing rustling of stones on these vast slopes.[18](5–6)

No distinctions are made between body and mind, between landscape and sensations: the upper valley is an all-encompassing environment, where the body, feeling threatened, becomes hyper-vigilant:

> And when one stands still, a moment only, in order to catch one's breath, one first thinks one hears one's own rapid heartbeat. But that has already

> fallen silent, and what remains to be heard—now distinctly, without a doubt—is the rustling of stones on the slopes.[19](7)

The physical distress enhanced by altitude isn't distinct from the unease caused by a formidable environment. The valley, where rocks and stones are the most conspicuous features, isn't a place of silence or quietness: it's a world of sound and fury. Nature, under its mineral form, doesn't lead the narrator toward a sense of abstraction or infinity but to the unsettling discovery of his body—not as a gendered body but as a living one, which belongs to this ungraspable environment with its own sound and fury.

Over the course of the novel, the narrator experiences a variety of physical symptoms typical of mountain sickness. In addition to the shortness of breath and arrhythmia described above, he suffers from insomnia, hyperesthesia, night sweats and terrors, and exhaustion. While describing his daily hikes, he exclaims: "But I discover terrible fatigues"[20] (121). These symptoms have also been associated with trauma and could be read as the result of a previous condition, either related to the loss of the loved one alluded to by the narrator, or even older. In fact, Schwarzenbach critic Decock reads *Das glückliche Tal* as the representation of an animist landscape unveiling the repressed trauma of the death of the narrator's lover (143). But I believe that this interpretation narrows the scope of the text. First, seeing the natural element of the landscape as symbol denies their materiality and that of the body surviving in their midst. Second, reading once again Schwarzenbach as the victim of an impossible (lesbian) love drama forecloses other dimensions, such as the very disorientation the narrator expresses. The strength and the beauty of the novel stem in fact from the undecidability of the symptoms and the distress related by the narrator: is it trauma, recent or ancient, is it mental illness, is it gender dysphoria, or is it simply the human condition? In the upper valley, the narrator is stranded in a state of meditation, where memories and introspection are intertwined with feelings, sensations, and descriptions of the surrounding nature. Transposed in the high-altitude environment, the moral pain and physical discomfort are made one, are made alive, tangible—and bearable.

The episode of the "happy valley" is described in several of Schwarzenbach's texts. But, in spite of its diary-like form, the novel discussed here was written at the Swiss clinic of Yverdon. Back in Europe in February 1938, Schwarzenbach had been undergoing yet another treatment for drug addiction and "schizoid personality." *Das glückliche Tal*, recalling events that took place three years before, has thus a double temporality: it narrates not one, but two crises. A letter to her friend Klaus Mann describes Schwarzenbach's state of mind in Yverdon:

> [A]fter three and a half months I am still in Yverdon, too, a horribly desolate place, even though I certainly don't have the need to be here. I am gradually turning crazy: I have come as far as closing the curtains and stopping up my ears with cotton balls, and start crying when a gentle sister

> disturbs me. Of course the clinic is surprised by this kind of "patient" I only weigh 49 kilos, sleep little, and take care not to observe any of the rules of the house.[21]
>
> (Fleischmann 1998, 172–3)

Like the valley, the clinic is "a horribly desolate place," a barely livable place. The hypersensitivity described in the letter reveals the same state of physical and sensorial distress as the one depicted in the novella. But the latter, in inscribing it in the milieu of the Lar Valley, brings it to life. The narrator's position of inadequacy finds its ecology, and its materiality, in a natural setting. This is, of course, a paradoxical and dysfunctional ecology, defined by the impossibility of finding a balance. The Lar Valley is a barely breathable space, and yet the narrator breathes in it: "we do live here, and we transfolk are home/not home in nature" (Stryker, Foreword, p. xvi, this volume). The sound of falling rocks is heartbreaking, yet the majesty of the surrounding cliffs is enchanting. The altitude's environment is a frontier space, a space above limits, where the narrator, threatened by nature itself, must constantly negotiate within its own bodily limitations:

> But resistance is of no use. I cannot silence the wind or stop the river, I cannot escape the tent walls nor this valley, and I cannot save myself a single hour.
>
> I am waiting[22] (46)

Waiting, here, is nothing but an attempt to exist, to inhabit what is thoroughly inhospitable—that is, one's body.

Frontier is the heart

There is no denying the orientalist dimension of Schwarzenbach's texts. However, the recurrence of high-altitude environments can be traced to the familiar as much as to the exotic. She grew up surrounded by mountains and took a lifelong interest in them. Like many characters in her novels, she was an accomplished skier and mountaineer. Her biography of the alpinist Lorenz Saladin (2007) shows, through detailed accounts of his expeditions, her knowledge of high-altitude environments and their dangers. Mountains are, for Schwarzenbach, a familiar space.

In Schwarzenbach's literary geography, the valley is a recurring pattern. It represents an ambivalent narrative ecosystem, triggering two seemingly contrary themes: isolation and circulation, reclusion from the world and openness to new worlds, remoteness and accessibility. In *Tod in Persia*, another novella recalling the same events, the Lar Valley is described as the place "where all ends." But in the opening of *Das glückliche Tal* it is precisely a place of beginnings:

> But the Lar Valley doesn't end there, far from it; do we even know where it leads?–Down to Mazandaran, into the Devil's Country at the Caspian Sea,

> the nomads say. … There jungles, rainforests, rice paddies, water buffaloes on melancholy dunes, humidity, malaria reign. In Gilan, the neighboring province to the West, the rice paddies are being drained on the Shah's orders, and the Chinese teach the malaria-farmers the difficult art of cultivating tea. … To the East the steppes begin, grazing grounds of the Turkmen. … At the harbor of Krasnovodsk begins the Russian railroad.[23] (8–9)

In this intuitive cartography, the Upper Lar Valley appears as the center of an ancient, quasi-mythical world that spirals strangely around it, as if the river itself, through its meanders, could lead to the Russian steppes. This intensive net of geographical connections seems to oppose the feeling of absolute loneliness that dominates in the novel. In fact, Schwarzenbach creates an experimental place, a frontier space where social structures are inapplicable, but where the border, the margin, becomes the heart. Triggered by the valley, the narrator's meditations travel from the local to the global, not in our postmodern sense of a small but alienated global village, but of a familiar infinity. The valley is where the world is ending, but it is also its heart, a place from where it can be contemplated, in the silence of the nights: "And I am looking–, deep in contemplation, painless silence–, and I hear the movement of the spheres"[24] (124). Again, this contemplation isn't set in an abstract void but can be traced to an actual environment, with its echoes, its rocks, and its cliffs–an ecology. The nocturnal "spheres," which evoke by association a bright starry night as it would appear from a remote and elevated area, are the nocturnal equivalent of the rustling stones, triggered in daylight by heat and animal activities.

A place of wonder, and disturbingly so, the valley is above all a place of beauty:

> [B]ut every morning when I leave my tent, I am surprised by the reborn beauty of this valley. Twilight still reigns, the night's lamp only just gone out, the world lies in the light sleep of the heights. There is no wind. The rock-crowned heads of the mountains touch the sky.[25] (47)

The tent, a protective shell, forms a porous skin between the body and its natural environment. Both belong to the same ecology, share the same rhythms of sleep and awakening. The narrator, like the personified mountains, touches the sky; that is, feels connected to the surrounding atmosphere, and grounded in it:

> My naked feet in the grass; its coolness touches me; sleep and dreamy warmth glide from my shoulders like an unwanted coat. I am unhurt, light, free—no pain touches me.[26] (47)

The narrator's sensations (bare feet on the cool grass) are inducing feelings (getting rid of the burden of an anxious sleep) in a deeply material way: body and mind are freed from pain through the connection with natural elements. The symmetry of the repetition ("its coolness touches me"; "no pain touches me") reveals the curative effect of the valley: sensations born through bodily contact with natural elements replace the pain brought on by human interaction—feelings of guilt, angst, inadequacy. The peace brought by the proximity with the sky also stems from "touching," connection, and material presence; again, infinity is not perceived as an abstraction but as a possibility.

As the valley and its inhabitants, human and animal, come back to life, set in motion by the warming morning sun, a cycle begins again: the valley is a door, an opening towards the rest of the world, toward adventure, toward hope, toward life. The peace and well-being born from the valley wake the desire to climb toward the summit, which in turn opens the paths toward other valleys, other cities, other worlds:

> And I'm still looking: uncountable life! … Ah, to climb higher! To look down from the roof of the world, on its mountain ranges and its abyss! Up to the blue Persian Gulf, belted by deserts! The sun is high, it is still summer, the heat still vibrates on the plain of Teheran, it is still cool in Schimran's green gardens—it is still time! Eagerness to follow the streets, the white tracks, the rivers—eagerness to discover the cities, the oases, the golden domes over palm trees—oh, unquenchable thirst![27] (51)

Having gained his footing on the top of the world, the narrator of *Das glückliche Tal* is ready to travel down again: not to be part of the world, but to wander, explore, write, dream, desire, and admire. In Schwarzenbach's own transecology, the only milieu to which the trans body and mind can belong is the ecology of the heights, the barely breathable, never comfortable, always mind-blowing, high-altitude environment, which echoes its own unsettling contradictions.

Notes

1 See, for instance, Joan Roughgarden, *The Genial Gene* (2009). Roughgarden offers a critique of the Darwinian notion of sexual selection and of its uses in contemporary biology.
2 Damit ich gesund werde, brachten sie mich in diesem Tal." All translations are mine.
3 "[D]as Œuvre steht ganz und gar im Schatten der Biografie, die allerdings faszinierend genug ist und Stoff für Filme, Bühne und Roman geliefert hat. Typisch ist jener rasante Satz, der sich in *Cosmopolitan* unter der Überschrift 'Leben auf der Überholspur' findet und der ein ganzes Leben zusammenfassen möchte: 'Sie war hochbegabt, reich, rastlos, leidenschaftlich, liebte Erika Mann und starb mit 34 Jahren.' (Anonym 2008)."
4 "Pour Annemarie Schwarzenbach, en la remerciant de promener sur cette terre son beau visage d'ange inconsolable."
5 "Die Annemarie hatte eben für mich damals diese Wirkung, die sie auf alle Menschen hatte: diese merkwürdige Mischung aus Mann und Frau. … Für mich … sah sie eben aus wie ich mir den Erzengel Gabriel im Paradies vorgestellt habe."
6 "[W]ie man einen jüngeren Bruder liebt."
7 Ludwig Binswanger's work was inspirational to the young Michel Foucault who prefaced and translated into French his article "Dream and Existence" in 1952.

8 The association between diagnoses of gender dysphoria, a term that hadn't been coined at the time, and schizoid personality disorders, was common in the 1930s and persisted until the end of the twentieth century.

9 "Die Schweiz ist ein kleines Land, und wenn ich unverheiratet hier bleibe … so wäre immer die Sorge, der Unwille, der Hass der Fronten, die Sensationslust der "Society" bliebe auf mich gerichtet."

10 "Soviel ich weiss gibt es in die Natur/Kraft bloss und ihrer Widerstand, nichts drittes."

11 See Fournier, Mat. 2014. "Lines of Flight.". *Transgender Studies Quarterly* 112. Durham, NC: Duke University Press.

12 "Schwarzenbachs Orienttexte lassen sich also in ihrer Gesamtheit als ein hochgradig intertextuelles Netzwerk beschreiben, wobei sich die Texte aufgrund der Verflechtung auf der Ebene des Raums und der Figuren gegenseitig erhellen können."

13 "Dieses Scheitern ist auch in ihrem Text erkennbar, in dem sie den Raum des absolut Anderen, als das sich Persien und seine Sitten der Reisenden präsentieren, sprachlich stets mit dem Vokabular des Europäers–und somit auch dem Kolonialisten—behaftet."

14 "Ich verstehe nichts von Hügeln, nichts von diesem Land. Es ist alles so schwierig."

15 "Landkarten trügen. Sie kennen nur *einen* Aspekt, und im Kreuz von Norden, Süden, Osten, Westen, bleibt der Demawend immer ein und derselbe. Ich habe aber einen anderen Demawend gesehen."

16 "Persepolis lag am Ende einer neuen Ebene, seine Säulen ragten auf hoher Terrasse wunderbar in den bewölkten Nachthimmel, und der Name wurde Wirklichkeit."

17 "Wir blieben den ganzen Tag in dem grossen Garten eines Englishen Freundes und dachten an Europa."

18 "Steht man irgendwo in der Mitte einer solchen Halde … dann kann man deutlich das unaufhörliche Rieseln des Gerölls hören. Dieses monotone, sehr leise Rieseln ist das einzige Geräusch in der Einöde …. Ich kenne kein unerträglicheres Geräuchs als das nie versiegende Rieseln der grossen Halden."

19 "Und bleibt man da stehen, einen Augenblick nur, um Atem zu schöpfen, dann meint man zuerst sein eigenes, rasch schlagendes Herz zu hören. Aber das ist schon verstummt, und was man immer noch hört—jetzt deutlich unmissverständlich–das sind die rieselnden Halden."

20 "Aber ich lerne furchtbare Ermündungen kennen."

21 "[A]auch bin ich, nach dreieinhalb Monaten, immer noch in Yverdon, einem fürchterlich öden Ort, obwohl ich es ganz gewiss nicht nötig habe."

> Ich schreibe an einem sonderbare Werkchen, und ich habe in meinem Leben noch nie so angestrengt gearbeitet. … Allmählich werde ich auch närrisch dabei: jetzt schlisse ich schon die Vorhänge, verstopfe die Ohren mit Watte, und weine wenn mich eine sanfte Schwester stört. Natürlich ist die Klinik ob solche "Patienten" verwundert. Ich wiege nur noch 49 Kilo, schlafe wenig, halte mich gründlichst an keinen Hausordnungen.

22 "Aber es ist nutzlos, sich aufzulehnen. Ich kann den Wind nicht zum Schweigen bringen und den Fluss nicht aufhalten, ich kann den Zeltwänden und diesem Tal nicht entgehen, ich kann mir keine einzige Stunde ersparen. Ich warte."

23 "Aber das Lahr-Tal ist damit noch längst nicht zu Ende; wissen wir überhaupt, wohin es führt?—Hinunter nach Mazanderan, in das Teufelsland am Kaspischen Meer, sagen die Nomaden. … Dort herrschen Dschungel, Urwald, Reisfelder, Wasserbüffel auf melancholischen Dünen, Feuchtigkeit, Malaria. In Gilan, der westlichen Nachbarprovinz, werden die Reisfelder auf Befehl des Schahs trockengelegt, und Chinesen lehren den Malaria-rn die schwierige Kunst der Teekultur. … Im Osten beginnen die Steppen, Weideplätze der Pendinischen und Theke-Turkmenen. … Im Hafen Krasnovodsk beginnt der russische Bahn."

24 "Und ich schaue-, Versunkenheit, schmerzlose Stille-, und ich höre die Sphären kreisen."

25 "[A]aber jeden Morgen, wenn ich mein Zelt verlasse, bin ich erstaunt über die wiedergeborene Schönheit dieses Tales. Noch herrscht Dämmerung, die Lampe der Nacht kaum erloschen, die Welt liegt in leichtem Höhenschlaf. Kein Wind regt sich. Die felsgekrönten Häupter der Berge berühren den Himmel."

26 "Meine nackten Füsse im Gras, Frische rührt mich an, Schlaf und Traumwärme gleiten von den Schultern wie ein lästiger Mantel. Ich bin unverletzt, leicht, frei—kein Schmerz rührt mich an."

27 "Ich schaue noch: vielfältiges Leben. … Ach hinaufzusteigen! Zu schauen vom Dach der Welt, über seine Randgebirge und Abstürze! Bis hinunter zur Bläue des Persichen Golfes, der von Wüsten umgürtet ist! Die Sonne steht hoch, noch ist es Sommer, noch zittert Hitze über die Ebene von Teheran, noch ist es kühl in den grünen Gärten von Schimran—noch ist es Zeit! Begierde, den Strassen zu folgen, den weissen Spuren, den Flüssen—Begierde, die Städte zu entdecken, Oasen, die goldenen Dome über Palmen—oh, unstillbarer Durst!"

References

Ahmed, Sara. 2006. *Queer Phenomenology: Orientations, Objects, Others*. Durham and London: Duke University Press.

Breslauer, Marianne and Kurt Wanner. 1997. *Wo ich mich leichter fühle als anderswo: Annemarie Schwarzenbach und ihrer Zeit in Graubünden*. Chur: Bündner Monatsblatt.

Bonstein, Carol. (2000). *Une Suisse rebelle* (documentary film). Geneva: Troubadour Films.

Decock, Sofie. 2010. *Papierfähnchen auf einer imaginäre Weltkarte*. Bielefeld: Aisthesis.

Decock, Sofie and Uta Schaffers. 2008. "Einleitung. Neue Wege zu Annemarie Schwarzenbach." In *Inside Out, Textorientierte Erkundungen des Werks von Annemarie Schwarzenbach*, edited by Sofie Decock and Uta Schaffers, 7–26. Bielefeld: Aisthesis.

Euchner, Maria. 2008. "Krankheit als Orientalismus in Schwarzenbachs Tod in Persien." In *Inside Out, Textorientierte Erkundungen des Werks von Annemarie Schwarzenbach*, edited by Sofie Decock and Uta Schaffers, 131–152. Bielefeld: Aisthesis.

Fähnders, Walter. 2010. "Zwischen Biografik und Werkanalyse: Die Schwarzenbach-Rezeption seit den 90er Jahren." In *Annemarie Schwarzenbach: Werk, Wirkung, Kontext*, edited by Mirella Carbone, 19–44. Bielefeld: Aisthesis.

Fleischmann, Uta. 1998. *Wir werden es schon zuwege bringen, das Leben*. Herbolzheim: Centaurus.

Halberstam, Judith. 1998. *Female Masculinities*. Durham and London: Duke University Press.

Kleist, Heinrich von. 1998. *Penthesilea*. Munich: Deutscher Taschenbuch.

Martin du Gard, Roger. 1931. *Confidences africaines*. Paris: Gallimard.

Rohlf, Sabine. 2010. "Neue Frauen und feminine Dichter—Annemarie Schwarzenbachs Figuren im Spannungsfeld zeitgenössischer Geschlechterkonstruktionen." In *Annemarie Schwarzenbach: Werk, Wirkung, Kontext*, edited by Mirella Carbone, 165–88. Bielefeld: Aisthesis.

Roughgarden, Joan. 2009. *The Genial Gene: Deconstructing Darwinian Selfishness*. Berkeley: University of California Press.

Schwarzenbach, Alexis. 2008. *Auf der Schwelle des Fremden: Das Leben der Annemarie Schwarzenbach*. Munich: Heyne.

Schwarzenbach, Alexis. 2007. *Maman, tu dois lire mon livre; Annemarie Schwarzenbach, sa mère et sa grand-mère*. Geneva: Metropolis.
Schwarzenbach, Annemarie. 2005. *Flucht nach oben*. Basel: Lenos.
Schwarzenbach, Annemarie. 2006. *Das glückliche Tal*. Basel: Lenos.
Schwarzenbach, Annemarie. 2007. *Lorenz Saladin: Ein Leben für die Berge*. Basel: Lenos.
Schwarzenbach, Annemarie. 2008a. *Bei diesem Regen*. Basel: Lenos.
Schwarzenbach, Annemarie. 2008b. *Eine Frau zu Sehen*. Zürich: Kein & Aber.
Schwarzenbach, Annemarie. 2008c. *Tod in Persien*. Basel: Lenos.
Schwarzenbach, Annemarie. 2008d. *Winter in Vorderasien: Tagebuch einer Reise*. Basel: Lenos.
Sedgwick, Eve Kosofsky. 2008. *Epistemology of the Closet*. Berkeley: University of California Press.

7 Transplacement

Nature and place in Carter Sickels' "Saving" and "Bittersweet"

Katie Hogan

Trans/rural literature[1]

Carter Sickels' beautifully written environmental novel, *The Evening Hour* (2012), illustrates a deep commitment to environmental justice, working-class struggles, and the close-knit connectedness of rural communities and families. Set in contemporary West Virginia, the novel's central concerns of mountaintop removal and rampant drug addiction are delineated in heart-breaking detail. Sickels shows how the landscape and the people are forever altered as they experience the slow destruction of their already fragile rural community. Carefully interwoven throughout the narrative is the issue of the main character's same-sex desire, furtively acted upon in West Virginia, where God, the American flag, coal, voting Republican, and traditional family reign.[2] *The Evening Hour* has garnered well-deserved critical praise for its portrait of stigmatized rural queer sexuality and environmental justice struggles; currently, the novel is being made into a film.

Although Sickels identifies as trans, *The Evening Hour* doesn't explicitly engage transgender perspectives on nature and environment, which is the focus of this chapter. Sickels' two briefer pieces, the short story "Saving," and the autobiographical essay "Bittersweet: On Transitioning and Finding Home," do. Both portray the complex experiences of transgender people as "outsiders" who are attached to rural place—a theme not explored in *The Evening Hour* or in much queer literature, activism, or scholarship.[3]

In these two shorter works, Sickels investigates connections among Appalachia, the environment, and trans attachment to rural nature and home, creating a nuanced and layered portrait of loyalty and longing for family, belonging, and rural existence. Challenging the typical portrait of the rural as wretched space—a view routinely found in queer studies, politics, and culture Sickels complicates the notion of the countryside as solely backward, misguided, and life-threatening. Instead, his trans literature shows that rural life is a place for trans/queers while he also carefully recognizes its many dangers.[4] In effect, he shows how rural America is a difficult place for queer people but it's still home. As Appalachian writer Melissa Range puts it, "I can't pretend I feel that I fit in back home, but I also can't pretend I feel quite at home anywhere else I go, either" (2015, 117).

Sickels captures the profound contradiction of both belonging and not belonging with imagination, vividly drawing transgender rural attachments, histories, and cultures. "Saving" and "Bittersweet" convey deep connections to rural place while laying out various toxicities—environmental poisoning and the social and political conservativism unique to small rural towns and farms. But Sickels also shows how reconnecting with working-class mountain people helped him return to his rural roots, and to begin to take steps toward integrating the threads of his trans and rural identities. Even with the alienation and violence that many queer people experience in the countryside, Sickels illustrates how trans people are attached to their often "inhospitable" native lands and homes. In presenting this intricate picture, he mines his own small-town rural and Appalachian background, taking us back to when he was read as a girl and spent a lot of time with his grandparents and large extended family. Sickels' goal is to show how the painful process of hiding and disclosing his queer identity at home never changes his belief about the power of rural place.

Rural queer and trans studies

In Lucas Crawford's analysis of trans spatial issues, Crawford states that "our environments move us as much as we move through them" (2013, 480). In other words, place is "more than a site of reaction," as Stephen Kuusisto (2015, xi) asserts; it's a complex amalgamation of knowledge, power, history, memories, and love. In this way, Sickels' literary contributions offer a unique understanding of rural experiences of place and home, an achievement that profoundly resonates with the mission of the relatively new field of rural queer studies. John Howard, one of the pioneers of this innovative research endeavor, has, in focusing on the problems engendered by "a hegemonic urban gay provincialism" (2016, 309), explored literary writing as a method for capturing the elusive complexities of rural queer experience. He states, "Queer fiction writers have long known" that rural space is not comprised of a "uniform, timeless hostility to queers" (2007, 102). As an example, Howard points out the quiet and casual acceptance of two gay men in Tennessee Williams' *Cat on a Hot Tin Roof* (2007, 102). Carter Sickels' literary fiction (and non-fiction) belongs to this tradition.

The "rural turn" in queer studies, although less focused on rural trans experience than on cisnormative queer experience, has demonstrated that geography is a central dimension in the analysis of gender and sexuality. This scholarship aims to do more than provide a mechanistic corrective to urban bias. Instead, it expands our understanding of a queer sense of place and challenges what Crawford calls queer urban "temporal superiority" (2017, 909). The field continues to loosen the grip of metro-chauvinism and questions the presumption that the city is the sole source of creativity and security for queer and trans people. Supposedly neutral concepts such as "the closet" and being "out," coupled with the deeply entrenched characterization of the rural as tragic, have ignored rural cultures and narrowed what queer studies can know about the vast rural worlds in the United States—and beyond.[5] Metro-chauvinism can even prevent scholars from challenging their own assumptions, contracting and constricting our insights on queer life, culture, and representation.

Not surprisingly, rural queer and trans studies serve as a productive lens through which to interpret Sickels' work. A capacious writer with a deep interest in place, Sickels inadvertently joins forces with trans and queer rural scholars in compelling and important ways. Neither peddles a racist, unthinking anti-urbanism, and both draw upon different discourses and methods to document rural trans and queer ecohistories, expanding our understanding of queer and trans identity in relation to place.[6] Sickels' work, like rural queer and trans scholarship, seeks to broaden the narratives we can write and tell.

The rural wasteland

Closeted. Dangerous. Dead-end existence. These words are frequently used to describe rural queer and trans life, conveying the entrenched idea that rural space has nothing to offer queer and trans people except isolation, violence, and death. Contrary to this popular belief, violence is not more prevalent in rural areas. In a critical analysis of the film *Brokeback Mountain*, John Howard points out that "hate crime statistics confirm" that queer bashers "tend to find their prey in the congested urban centers. The pistol-whipping of Matthew Shepard may lend itself to high drama, notably *The Laramie Project*, but the real dangers are on the mean streets. Think *Torch Song Trilogy*" (Howard 2007, 102). Howard's analysis further develops the point that metro-chauvinism plays a role in distorting our understanding of rural violence. In the film, Jack proposes that he and Ennis, two cowboy lovers who have sex on the mountain, should live together as romantic partners. Ennis rejects the plan outright, saying that they would be putting themselves "in the wrong place" and would soon "be dead." Hopeless, Jack travels from Wyoming to Mexico for tourist gay sex—even though sexual opportunities await men who have sex with men at the many truck stops and rest areas along the way. Howard believes the film embodies a "smug urban condescension" that "displace[s] homophobic violence onto the hinterlands" (2007, 102). He also points out that the film lacks any rural queer historical knowledge and recycles the most entrenched stereotypes of queer men by presenting queer sex and love as under constant threat and untenable.

Stina Soderling and Carter Sickels' work meshes with these notions about rural violence that circulate in US culture. Soderling states, "The rural is approached with fear and presented as an Other at constant risk of demise. Those who are stupid enough to not leave for the city are to blame for their own death" (2016, 343).[7] And Sickels challenges the "natural" linkage between rural queers and violence:

> For so long, the only narrative we hear about concerning queers living in rural areas is one of violence—and while I think that's an important story to tell, it's not the only one. I'm interested in stories that break open that

> simple binary—that cities are safe and rural spaces are unsafe for LGBTQ folks—and explore the many ways one can be queer and exist in rural or natural spaces.
>
> (Carter Sickels, email message to author, June 21, 2016)

Rural queer studies emerged to refute these ideas of the rural as inevitably and naturally violent, and it has consistently challenged the idea that the rural is an unlivable space for queers. It has also introduced the theoretical concept of metronormativity, which refers to an urban bias in queer culture (including commercial culture), politics, and academic theory.[8]

Metronormativity consistently elevates the urban over the rural by way of a rigid binary separation that misrepresents, and often actively belittles, rural cultures. It casts the countryside as a pathetic, inhospitable closet—stuffed with suffering rural queer and trans people—and tells rural queers that they have no choice but to "Get Thee to a City," as anthropologist Kath Weston (1995) captured the message in her famous essay on the city and the gay imaginary.[9] In contrast to this formidable belief system, rural queer studies contests metronormativity as a way to reveal the complexity and heterogeneity of rural queer space. Scholars argue that rural queers without the economic means to relocate to large, expensive cities, such as New York, San Francisco, Boston, or Chicago, and those without an interest in urban life, are worthy of critical and creative attention.

John Howard's highly regarded book, *Men Like That: A Southern Queer History* (1999), is a touchstone in the field.[10] A Mississippi native, Howard is well aware that his home state is considered one of the least progressive, most "backward" states in the United States. He nonetheless complicates this idea by documenting a striking network of male queer/trans rural cultures in the first half of the twentieth century. He makes the case that, despite significant peril, many queer Mississippians invented, and participated in, rural queer and transsexual networks with resilience, strength, and enthusiasm.[11] While Howard acknowledges that many queer Mississippians followed—and still follow—the classical trajectory of rural to urban migration, he repeatedly asserts that many queers—even those with economic means—did not leave the state and that their lives amounted to more than "making do" in a rural wasteland. One of his interviewees, a trans woman named Rickie Lee Smith, "chose to live nowhere else but Mississippi" (Howard 1999, 124).

Despite the impressive work rural queer studies is doing, metro-chauvinism still holds sway, and rural queer and trans experience remains understudied.[12] Will Fellows' book, *Farm Boys* (1996), includes testimony by Midwestern gay men who feel happier and freer in urban centers, but it also showcases respondents who feel dislocated in cities. These "farm boys" internalize the urban "get out of the sticks" message and "escape" to the city; but once they relocate, they often experience even more isolation and loneliness than they did at home. Katherine Schweighofer (2016) calls this phenomenon "double alienation," a term that refers to "rural-raised" queer people who feel they don't belong in the countryside but also feel "equally out of place" in the city (234). Fellows explains that, "For many [farm boys], the dislocation of living in an urban culture after growing up rural was in some ways similar to that of being gay but living in a heterosexist culture; in both regards, they felt like outsiders" (Fellows 1996, 311).

Nonetheless, *Farm Boys* offers more than stories of rural alienation and urban disappointment. Several interviewees talk openly about a sense of peace and freedom they derived from their experience in rural nature. Former farm boy Henry Bauer states, "I was in heaven when I was in the woods; it was an escape" (Fellows 1996, 67). In fact, to many of Fellows' interviewees, nature and rural place assisted queer sexual activity, suggesting what Colin Johnson calls a "pragmatics of pleasure"(2013, 1), a term that explains the abundant same-sex sexual activity in rural America despite the absence of an urban-centric model of gay identity and gay culture.[13] In addition, as *Farm Boys* and other rural queer research shows, some rural queers give up on city life and return to their home communities where they try to find ways to practice their queer sexualities and genders in the context of rural culture. Poet Aaron Smith, a queer West Virginia native, sums it up as follows: "It's wanting to get out. It's getting out. It's going back" (2015, 197). This "reverse migration pattern" complicates the presumption about city life as the only reputable choice. Annes and Redlin (2012) argue that, while the city has a liberating effect on rural gay men, it is not always a final destination.[14]

From the perspective of queer metronormativity, those queers who stay on farms or in rural small towns and those who migrate and then return are viewed as unsophisticated and conservative, and mostly written off as frightened conformists whose rural queer attachments are perplexing and pitiful. As Mary Gray explains, this view exists because "rurality itself is depicted as antithetical to LGBT identities" (2009, 12). This still-entrenched belief reaffirms the notion that what counts as "out" and visible reflects an urban ideal that renders all rural queer and trans people as closeted and weak.[15]

Not only do rural individuals with queer sexualities and genders have to negotiate the contradiction of simultaneously belonging and not belonging—being "out of place"—in both urban and rural contexts; they must also contend with being judged according to a "universal" urban standard that is not truly universal. The closet model ignores how rural queers have their own "strategies of recognition," a term coined by Gray that refers to the myriad ways of being visible/out beyond the one hegemonic metro-chauvinistic paradigm (2009, 168). In an article that explores fictional queer lives that resist metronormativity, Mary Pat Brady explains, "Visibility and the closet … are inadequate metaphors for approaching an understanding of nonurban queer cultures" (2016, 120).[16] The closet metaphor is also highly incompatible with transgender people's strategies of recognition; to many trans people queer metro-chauvinism can seem haughty, self-righteous, and simplistic. In *Trans/Portraits* (Shultz 2015), interviewees mention their preference for the word "disclose" instead of the phrase "coming out" to describe their approach to visibility. One respondent also explains that, because transgender people are watched and stared at in public, treated as oddities and spectacles, they often don't find queer pride parades appealing or empowering (Olivia 2015, 83). Relatedly, Lucas Crawford casts rural modes of "'passing' as a void notion in a small town where everyone knows you" and argues that rural trans illegibility has analytic power: "I wonder if the way in which rural people are largely ignored by urban queer theory lets us experience something more exhilarating than passing: imperceptibility" (2013, 481). For Crawford, "imperceptibility is something entirely different from being 'in the closet,' as rural queers are so often read" (2013, 481). Instead, "imperceptibility and lack of recognition" make complex "rural styles of transgender" lives possible (2013, 481). This trans spatial critique compounds the need for a paradigm that accounts for many ways of being queer and trans instead of the one dominant urban-centric closet/out model.

Rural scholar Kelly Baker offers an alternative paradigm that accounts for variation in visibility and "outness"/"disclosure" in rural and urban contexts. An advantage of this model is that it creates an opportunity to explore rural and urban contexts in a less judgmental way and suggests how rural existence energizes queers—a perspective woefully absent in theory, politics, and culture. According to Baker, urban visibility politics is based on a "different-but-equal paradigm," whereas rural visibility politics is based on a "different-but-similar" paradigm (2016, 42). From a rural perspective, this would mean that loyalty to community is not necessarily a manifestation of assimilationist, conservative ideology (2016, 42). Mary Gray explains, "one's reputation as a familiar local is valued above all other identity claims" (2009, 31). Baker elaborates, saying, "many rural LGBT people find that they are judged primarily by their farming abilities, their community involvement, and their roles as good neighbors" (2016, 38–9). Mark Hain adds to an analysis of rural strategies of recognition when he says, "[B]eing perceived as a 'good person' may be valued more than sexual difference is condemned" (2016, 172). Furthermore, in Baker's interviews with queer rural people, she found that many individuals treat their queer sexuality and queer gender as a thread of their identity, not a core element. When gender identity and queer sexuality are experienced as such, it becomes clear how the urban closet model loses its salience (Baker 2016, 41).

A dimension not addressed in the "different-but-similar"/"sameness" rural queer identity paradigm discussion is the hegemonic power of whiteness/white supremacy, which largely determines which rural trans and queer people can claim "different-but-similar/sameness" status in the American countryside. Miriam J. Abelson's "'You Aren't from Around Here': Race, Masculinity, and Rural Transgender Men" documents how "whiteness is often a key component of claiming sameness in predominately white rural places" (2016, 1537). Although rural studies and rural queer studies research make clear that rural space is not devoid of people of color, immigrants, and queers, many of these groups encounter racism, transphobia, homophobia, and xenophobia in small towns and rural areas. In my interview with René Pico, a college professor and cis gay man who was born and raised on a farm in Puerto Rico and now lives on a farm in rural Pennsylvania with his male partner, racism is a central dimension of his everyday rural life. Living approximately eight miles from "one of the biggest KKK clans in the region," René is routinely denied service at businesses and shops because Spanish is his first language. "I have been denied fair treatment as soon as I speak. I have to ask my partner to speak on my behalf for services" (René Pico, email message to author, November 18, 2016). Shifting away from a metro-chauvinistic model of the closet to one that considers queers in rural regions illuminates the deep attachments and harsh difficulties that trans and queer people face. With depth, grace, and specificity, Carter Sickels' trans literature offers a poignant elucidation of queer/trans rural worlds, illuminating the way space, place, and identity are entangled concepts and practices.

Looking homeward

Queer/trans people have been deemed "unnatural" and "against nature" for centuries, so it is fitting that conflicting emotions about living in the countryside are central to Carter Sickels' writing. Katherine Thorsteinson and Hee-Jung Serenity Joo (Chapter 2, this volume) also unpack the complexity of the "natural" by showing how it is used as a weapon against *Transparent*'s Maura Pfefferman, who experiences hatred and bigotry at two different "nature" camps. Maura's encounters with residents at these camps illuminate how she is perceived as both the "wrong *kind* of woman" and "the wrong *kind* of trans"—exemplifying an ingroup-outgroup mentality that is skillfully exposed and resisted in *Transparent*. Although not focused on denaturalizing nature or the complexities of seasonal nature-spaces as "safe havens" for trans people, Sickels' short story "Saving" and his personal essay "Bittersweet: On Transitioning and Finding Home" approach queer and trans rural experience as a potent mixture of conflicting elements. In Sickels' work, beauty and affection, hardship and violence clash, but the author deliberately depicts this collide without resorting to images of the countryside as a totalized wasteland on the one hand or as rife with "hillbilly nostalgia" on the other.[17] Rural sentimentality in particular is something Sickels explicitly rejects:

> In my writing, I wanted to express my love for the place, but I also did not want to hide from the darker, harder parts: the claustrophobic web of religion, the stark closed mindedness, the drug addiction and poverty and destroyed land and ruined waters.
>
> (2015, 77)

"Saving" and "Bittersweet" explore the "getting out" of the country myth that pervades discussions of rural space, and both pieces adopt a "homeward" gaze, allowing Sickels to spell out the significance of home and nature from a personal rural trans perspective.[18] While Sickels has mixed feelings about the rural as a viable place for him and his trans character Dean, he embraces, even celebrates, rural heritage at the same time that he critiques it.[19] Neither work depicts a trans individual as a permanent resident of Appalachia, but both convey a deep longing for home place; the tentative, cautious returns "back home" that Sickels and his character make are rendered with both ambivalence and affection.[20]

"Saving" sets up a rich tension between an urban-centered world, embodied by Jillian, and a rural, working-class world, embodied by Dean. Jillian, a white cisgender woman who grew up in a wealthy Long Island suburb, is a trendy experimental filmmaker who identifies as queer. Her family attends the opera, discusses cutting-edge art openings at dinner, and, because "Everyone in the family goes to therapy, they're always telling each other what they feel" (Sickels 2012, 34). Jillian met Dean at a party in Williamsburg, Brooklyn and felt an immediate attraction. In addition to being in a romantic/sexual relationship with him, she is making a film about Dean's gender transition.

In comparison, Dean grew up as "a country kid. ... Not boy or girl, just a kid," but was read as a girl by family and society. As an adult, he claimed a trans identity and eventually "escaped" to the city, a space that has brought him both freedom and alienation (Sickels 2012, 26). He explains that the city hasn't offered a solution: "I moved to the city to escape the isolation of my childhood, but it followed me, a disease in my bones" (2012, 36). Most of Dean's city friends are Jillian's, not his, and though Dean doesn't know it until later in the story, Jillian is betraying him by having an affair with another artist, who has a background similar to hers.

The story takes place in Perry, Kentucky, the rural mountainous coal town where Dean grew up. He's returned to see his grandmother, who is ailing in a nursing home, and to take care of her house and affairs. He had to convince Jillian to accompany him; at first, "she didn't want to leave the city" but she changed her mind once "she started seeing it as a filming opportunity. ... 'Trans guy in Appalachia' " (Sickels 2012, 22).

Dean's return to Perry and his grandmother's house stirs up painful memories. Dean's mother died of cancer when he was 8; his father, a violent alcoholic, committed suicide after her death, and his grandmother, who raised him, is suffering from dementia. His grandmother's ramshackle house has a sagging roof and old tires in the yard, and a "dented GE washer sits on the front porch" (Sickels 2012, 21). As her illness progressed she became a hoarder, and stuffed the house with newspapers, old bills, furniture, clothes, and knick-knacks (2012, 21). When Dean sees the house, he feels overwhelmed and suffocated. Nonetheless, he is drawn to this impoverished but beautiful town and place (2012, 26). He notices the stream in the backyard, the woods, trees, birds, and nature overtaking the old tires and aluminum cans, evoking the alignment he feels with nature, a theme that Elizabeth Parker explores in her nuanced analysis of gender and nature in *The Danish Girl* (Chapter 1, this volume). But when Dean suddenly sees the landscape and house from Jillian's point of view, he feels "embarrassed, her seeing where I grew up" (Sickels 2012, 22).

The story charts Dean and Jillian's reactions to Perry, offering the reader insight into the conflict between them, which is linked, in part, to the intractable queer urban/rural binary. Dean watches Jillian taking in the environment and thinks, "She looks beautiful, and out of place," seeing her as an outsider who will never understand him or Perry (Sickels 2012, 21). Soon after they are settled, Dean talks with Jillian about why they cannot disclose their queer relationship and his gender identity to Paul, the angry white, straight, cis male neighbor, or to the caretakers in the nursing home where his grandmother now lives. He warns that they cannot be "raging queers" in the Appalachian countryside. Jillian shoots back, "I'm not straight. That's not who I am," betraying an insistent queer urban visibility politics that the story contests (2012, 33).

This exchange between Dean and Jillian is significant, since it unwittingly echoes academic debates about queer theory, in which normativity is pitted against transgression. Jillian's demand that they be "out" in rural Perry shows her claim to rebellion over conformity, but this binary, like the urban/rural divide, ultimately occludes the complexities of rural gender and sexuality instead of delineating them (Hines 2010, 597, 600).

Trans scholar Petra Doan, who describes her spatial experience of gender "as a kind of moving target" rather than a binary, speaks to the complexity of trans geographies (2010, 645). Writing that moving through space as a trans woman can involve transformation, confrontation, or both, Doan explains that her gender performance is inspired by space, place, and onlookers:

> There are places in which I never raise my voice above a whisper, such as public restrooms. In addition, when I use public transportation in unfamiliar locations or when I travel the back roads through unfamiliar terrain, I rarely engage those around me in idle conversation until I am able to get a reading on how invested they might be in the dichotomy of gender. I am not shy, just careful. I recognize that my gender performance is simultaneously modulated by the observers of my gender as well as the spaces in which we interact. These modulations do not shift my own sense of gender, but they do shape the visibility and impact of my gender performance. Sometimes I can choose when to perform my gender in ways that might expand the boundaries of the gender dichotomy and sometimes I cannot.
>
> (2010, 648)

Similar to Doan, Dean tempers his gender performance. To him, gender comprises an inner experience, but it's also adjusted based on space and unknown observers—as Doan describes. Dean downplays this internal sense of gender from the neighbor and the staff at the nursing home, not because he's a conformist or assimilationist—as Jillian's angry response implies—but because he is actively negotiating trans geographies. In contrast, Jillian characterizes queer identification as fixed; wherever she goes, she is "queer" in the exact same way. In addition, Jillian may be more invested in seeing herself (and Dean) as monolithically transgressive because such an image resonates with her reputation as a "hip" filmmaker.

Here, and throughout the story, Dean's inner thoughts reveal his uneasiness about Jillian's demand for visibility and her loquaciousness, which clash with his more taciturn style: "I was an only child, didn't have many friends. Grandma and I did not need to talk about what was in our hearts, we weren't that kind" (Sickels 2012, 25). When they arrive at Dean's grandmother's house, Jillian begins pressing Dean for details of his childhood. She "wants to know everything. *Trans*parency, she says" (2012, 25, emphasis added). Given that she's secretly having an affair, her demand for truthfulness is ironic. But the larger point is that her insistence that Dean divulge his "trans" feelings—her doggedness to get more information on "Trans guy in Appalachia"—creates mistrust. When Dean shares that his father used to hit his mother, "Jillian's face cracked with interest, and I quickly backpedaled, downplaying it" (2012, 27); when Dean "hear[s] Jillian come in, I set the clothes aside, thinking she'll want to use them for the film" (2012, 41).

Before the visit to Perry, Dean believed that Jillian was the only person who really "saw" him. Now he is beginning to feel that she might not see him at all. He thinks, "She has shot me hundreds of times, but there is still so much she doesn't know" (Sickels 2012, 24). Dean predicts, when the film is completed, that he "will not recognize" himself (Sickels 2015, 50). These startling lines suggest how human identities and complexities are beyond our framings, but they also indicate that Jillian doesn't truly know Dean and doesn't want to. When they are in the city, the film project makes Dean feel seen; in the country, the project makes him feel used.

Being at his grandmother's house, Dean feels vulnerable and overcome with emotion. Jillian, in contrast, says, "I can't wait to start filming" (Sickels 2012, 22). Dean, his grandmother, and the home are her "material." Dean begins to think of Jillian's camera as "a weird monster eye" that is stalking him; when she brings it to the nursing home to film his grandmother, he thinks she is like "a kid caught stealing candy" (2012, 31, 32).[21] After Jillian confesses that she's having an affair, Dean responds that the only reason she came to Perry was "to make a goddam film" (2012, 45).

"Saving" does not end happily, but it also does not end tragically. Dean's grandmother is declining, although she has bursts of loving recognition, and Dean will sell their home to the coal company. His relationship with Jillian is over, which angers him more than it surprises him, and he accidently kills an abandoned puppy with his car. On the surface, no one seems to have been saved in "Saving," but more than any other character in the story Dean is the one who is saved, and his grandmother, who symbolizes the home place and the land, is doing the saving.

Dean returns to Perry ostensibly to help his grandmother, but on a deeper level the visit is about Dean "saving" himself. The "hiding" he engages in throughout the story—from the neighbor, at the nursing home, in Perry—is more than a public safety measure. It's not about his being "closeted," in the urban sense of the word, as it reflects the shame he feels for leaving home and not being there to take care of his grandmother. He cannot stop thinking about how his grandmother raised him and expected him to return to Perry: "I want to apologize to my grandmother, how sorry I am for leaving her" (Sickels 2012, 49).

Dean avoids the diner where his grandmother worked for years—he doesn't want to face community members who might see him as a traitor. He feels he has betrayed his grandmother and the rural culture from which he comes: "I could have come out to stay with her. I'm all she has" (2012, 31).

The themes of hiding and seeing that are central to this story resonate with rural queer and trans studies' critiques of the closet. Dean says, "My grandmother taught me to look at what was hidden, to see what was right in front of my eyes" (Sickels 2012, 51). His grandmother's lesson suggests that the hidden is not so "hidden" after all, that the assumption that someone is closeted might reflect the inability to see what is "right in front of [one's] eyes," a central theme of Sickels' story and rural queer and trans studies.

And it turns out that Dean's grandmother, on some intuitive level, knows about his feelings of guilt and his reluctance to disclose his gender identity. In the story's last line, she says, "You're a good boy. A real good boy," making clear that she not only forgives him for leaving, but also sees him for who he is (Sickels 2012, 51). The line merges her acceptance of his gender identity—"good *boy*"—with her forgiveness, releasing him from the shame he feels about not fulfilling his commitment to her and the socially imposed anxiety he feels because of his gender identity. In this last exchange, Dean's trans identity and his "country kid" self are integrated again, making clear how the grandmother and the home place "save" him.[22] Dean believes that life in Perry with his grandmother "is the deepest and oldest" part of his life (2012, 43).

Sickels' story emphasizes home, but it also focuses on rural nature as more than a side issue or simple backdrop. In the story, nature opens up opportunities for Dean to embrace the "rejected rural" and blend it with his trans identity, which generates an expansive, encompassing, flexible self—the very opposite of what is typically associated with trans people in rural space. Here, nature functions as solace, balm, and refuge for those who are exiled. As Sickels explains, "For many queer people, rural space, or nature, is a crucial part of who they are and how they exist in the world" (Carter Sickels, email message to author, June 21, 2016).[23]

Queer rural scholars Mary Pat Brady and Mark Hain also analyze the connection between rural space/nature and queer identity. In an article on nonurban fiction, Brady says, "[Characters'] comments on the areas around them are not mere asides, but rather engagements with the place as a way to understand themselves" (2016, 119). Hain, echoing Brady and Sickels, discusses how, in rural cultures, engagement with nature "is perceived as vital identity work" and that "nature allows space for … self-reflection and self-invention" (2016, 175, 176). Sociologist Sally Hines concurs: "[T]he language of identity is spatially strategic and dependent upon place and context" (2010, 600).

Throughout "Saving," Dean reminisces about the life he and his grandmother created, surrounded by mountains, trees, birds, streams, and fresh air. Looking back, he acknowledges the loss, isolation, generations of family violence and poverty, but always laces these memories with a fierce love for place and land. He recalls "camping in the Smoky Mountains" with his parents, thinking he "never wanted to leave" (Sickels 2012, 34). He remembers something his grandmother taught him: "Everything counts. Animals, trees. Everything's connected, the dead and living" (2012, 34).[24] Dean had to leave home to understand himself, but he reconnects to home and nature to be more fully who he is.[25]

"Bittersweet: On Transitioning and Finding Home" tracks the emotional distance between Carter Sickels and his family, a breach that began when he was a teenager and identified as a lesbian. An exile in his own family and community—"I did not fit in" (2015, 75)—Sickels says this feeling increased when he transitioned, but by that time he had already left home and visited only occasionally. He speaks repeatedly of the isolation and loneliness he experienced and still experiences. Nonetheless, these negative feelings never drown out his love of place and family, which continue to enrich his life. The silence and isolation of childhood never corrode his love for home.[26]

Visiting West Virginia to conduct research on mountaintop removal for *The Evening Hour*, Sickels says he was "overcome with a longing to return to [his] roots" (2015, 73). He recalls the painful and joyous memories of his childhood: the large family gatherings, particularly when he was a child, fused with the painful silence when his sexual and gender identity made him "less recognizable to my family" (2015, 75). The stifling conversations, allegiance to conservative Christianity and the Republican Party, and his sense of being an exile co-existed with deep attachment to place and home. Meeting West Virginia people, he says, "returned me to my roots" (2015, 77).

He does not disclose his gender identity to anyone during his research visit; he's there to understand the community's struggles against environmental poisoning, poverty, and addiction. At the same time, he is also protecting himself against violence. Trans people are killed in cities, but as a gay trans man in conservative West Virginia, he experiences a paralyzing fear each time he stops for gas or uses a rest-room. Nevertheless, Sickels feels at home, as if seeing his relatives and childhood in the landscape and people. In getting to know the community, he befriends a white, straight cisgender Vietnam veteran to whom Sickels sends a letter with the news of his first novel and gender transition. He receives a warm, accepting response a few days later, another instance in which Sickels' writing resists rural monoliths and stereotypes (2015, 79). As Sickels puts it, "I still write about this place of dreams and nightmares, my yearning mixed with fear" (2015, 79). Sickels approaches trans and queer rural experience without demonizing or idealizing it, offering a way to loosen the grip of queer metronormativity.

Beyond the urban/rural divide

In their introduction to *Queering the Countryside: New Frontiers in Rural Queer Studies*, Colin R. Johnson, Brian J. Gilley, and Mary Gray mention how queer studies scholars, though sympathetic to the claim that rural queer history and culture have been ignored, are nevertheless impatient with the repeated critique that they, and queer culture beyond academia, are relentlessly metronormative (2016, 9). Taking these scholars' arguments to heart, Johnson, Gilley, and Gray write that rural queer studies scholars must clearly articulate

> [W]hat is it, precisely, that queer studies scholars' metro-chauvinism has supposedly forestalled them from seeing, or understanding, about the predicament of queer life, or its potential.
>
> (Johnson, Gilley, and Gray 2016, 9)

Carter Sickels' work—similar to other artists and creative writers analyzed in the chapters of this volume—dismantles dualisms and, in doing so, offers a clear response to this question through his dynamic literary approach to queer and trans spatial relations attuned to environmental and spatial heterogeneity.[27] Instead of portraying urban space as The Promised Land and rural areas as The Wasteland, he, like many rural queer and trans studies scholars and writers, offers a more capacious and complex vision, one that neither eclipses nor exaggerates a particular space.[28] His work demonstrates how exploring queer/trans rural place is a significant, productive pursuit that contributes to the development of a trans literary canon and to rural queer/trans studies more generally. Less about how rural queers/trans experiences have been systematically left out, although that is a central part of the project, his work suggests how a more complex understanding of attachments to home and rural place can expand the field itself. He shows how "hiding" might not be "hiding"—or, at the very least, the reasons for one's "hiding" might be more complicated than traditional queer scholarship and culture have supposed. He demonstrates that rural areas are not sites of unmitigated hostility and backwardness, and that urban life has components of painful compromise that are mostly left unspoken in scholarship and culture. Compromise, trade-offs, difficulties, and serious challenges saturate the lives of urban queer and trans people alike, not just the lives of those in flyover states, the South, or in small towns and rural regions. Sickels shows that, to a certain degree, we are all attached to inhospitable places.[29]

Notes

1 The phrase "trans/rural literatures" is Crawford's (2016).

2 While this typical characterization of West Virginia is accurate, it does not foreclose the complex histories and presence of people of color, queers, Indigenous people, and progressives in the state. As Elizabeth Catte argues in *What You Are Getting Wrong About Appalachia*, "Many things about Appalachia may be true simultaneously" (2018, 52).

3 Lucas Crawford's article, "Transgender Without Organs? Mobilizing a Geo-affective Theory of Gender Modification" and, to a lesser extent, "A Good Ol' Country time: Does Queer Rural Temporality Exist?," specifically address trans rurality from a literary and queer/trans perspective.

4 Carter Sickels identifies as queer, a term he finds most inclusive of his trans and gay male identities. However, I sometimes use the terms trans or transgender to emphasize how "Saving" and "Bittersweet" focus specifically on rural trans experience.

5 In Johnson, Gilley, and Gray's introduction to *Queering the Countryside: New Frontiers in Rural Queer Studies*, the authors point out that, while urban space garners most of the scholarly attention in various fields,

> urban space is oddly atypical of the U.S. national landscape. Including Alaska, statistically urbanized areas account for a mere 2.62 percent of the United States' total land area. … This means in the United States, almost 97 percent of territory—some 3,443,773 square miles in total—remains nonurbanized, or "rural," in character.
>
> (2016, 1)

6 Sickels' narratives of returning to rural home place after moving to urban centers brings to mind bell hooks' work on the importance of creating African-American "ecohistories." hooks points out that modern African-American life is typically associated with cities—"little is shared about the agrarian lives of black folk" (2015, 14). African Americans have a lengthy ecohistory that connects them to rural spaces, particularly in the South. Much of that history involves widespread trauma: slavery, lynching, sharecropping, Jim Crow, systemic racism—all brutal experiences that could easily overwhelm one's memories of peace in nature. As hooks says, "nature, once seen as a freeing place, became a fearful place. That silence has kept us from knowing the ecohistories of black folks" (2015, 14–15). hooks contends that it is crucial for African Americans to reclaim their connection to nature, even if it means directly revisiting this violence and trauma. Sickels', and his characters', alienation from nature does not compare to the situation of African Americans, yet his work questions the deleterious effects of a metronormative mandate that automatically links queers to cities, reinforcing urban-centric narratives that wipe out, or distort, a rural trans/queer heritage. Because Sickels' work actively addresses attachments between rural nature and trans and queer people, he is helping create queer/trans ecohistories.

7 Soderling's analysis resonates with Lucas Crawford's point that queer and trans metronormativity construct the city as "where queers *do* queerness, and the country is where things *are done* to queers" (Crawford 2017, 917, emphasis in original).

8 Judith (Jack) Halberstam coined the term metronormativity (2005, 36–8).

9 As Johnson, Gilley, and Gray explain, "The structure of thinking that Weston reveals—an urban versus a rural dichotomy—is written throughout the history of queer studies as well as through the popular and activist queer imagination" (2016, 12). Weston's landmark essay helped pave the way for rural queer and trans studies.

10 In addition to Weston (1995) and Howard (1999, 2007, 2016), there are other rural queer studies scholars who have produced critical work: see Fellows (1996); Halberstam (2005); Gray (2009); Herring (2010); Johnson (2013); Keller (2015), and Gray, Johnson, and Gilley (2016). For a powerful recent response to stereotypical queer metronormative characterizations of the American South, see Holloway (2016).

11 As Mark Hain points out, it's not uncommon to encounter rural queer people who are "leading sometimes difficult but also fulfilling lives outside urban environments" (2016, 170). Jay Michaelson's insights on the underlying assumptions of the It Gets Better (IGB) campaign resonate strongly with Hain's observation:

> "It Gets Better" is an oversimplification. We queers know this; we know that sometimes it gets worse before it gets better, and that sometimes, it gets worse again, and that other times, some days you just have to get through one at a time. And yet, like our other metaphors—Coming Out, Transitioning—It Gets Better suggests a linearity that is at odds with LGBTQ experience, even as it also, helpfully, offers hope.
>
> (Michaelson 2012)

12 See Mark Hain's critique of the *It Gets Better* project. Hain looks at the messages in the project's videos and concludes that "part of getting better is getting out of the

narrow-minded rural area, the oppressive small town, the unenlightened 'fly over states,' because happiness, acceptance, self-fulfillment, and others like you are to be found only in coastal urban centers" (2016, 164–5). He suggests that the IGB campaign's message is that "there must be something wrong with the gay [or trans] person who does not migrate to the city" (2016, 165).

13 Fellows' informant Jim Cross supports Johnson's insight when he discusses how nature facilitates gay farm boy sex: "Our farm sat at the edge of a heavily wooded area, twelve to fifteen acres. Oftentimes we'd just hike into the woods. That was real safe, because there were lots of places where nobody would find you. We made a number of hideaway places. We'd get a bunch of leaves and make it like a bed" (1996, 78).

14 Annes and Redlin argue that, to the rural gay men in their study, the city offers a model of masculinity that does not mesh with their felt identities. In particular, they found that the rural gay men they interviewed feared the perceived "feminizing" effect of the city (2012, 56–7). One of Fellows' interviewees, Todd Ruther, speaks to this fear: "Some people assume that if you're gay you're going to move to the biggest city and wear the flamiest clothes and learn how to walk the swish. I would rather be able to go back to live in my hometown" (1996, 304).

15 What counts as "out" is also racialized. See Marlon B. Ross, "Beyond the Closet as Raceless Paradigm" (2005).

16 Katherine Schweighofer suggests that "the closet contains too many metronormative elements to function as a useful metaphor for rural gay and lesbian identity" (2016, 239). John Howard concurs, arguing that the urban-centric notion of a self-proclaimed gay identity fails to capture the same-sex practices and networks he documents; he concludes that "identity" is a concept that crudely simplifies rural queer lives, illuminating the "limitations of gay identity politics" (1999, 306). Howard's study of queer and trans Mississippians suggests that queer gender and queer sexuality do not constitute identity elements that make up a centered self; rather, gender and sexuality are performances and practices. For the queer rural Mississippians he studied, it's not so much that there were numerous ways to be "gay" but that there were numerous ways "to do" queer and gender desire. Howard's take resonates with Judith Butler's (1990) assertion that gender and sexuality are something you do, rather than something you are.

17 The term "hillbilly nostalgia" is Range's (2015, 117).

18 The "getting out" myth phrase comes from Tennessee Jones' "Getting Out: The Grief of Transformation" (2015, 121–34). Jones (akin to Sickels) uses literature to address the rural/urban binary: "I wrote myself out of Appalachia, my particular nexus of what had to be escaped and denied in order to survive, and I wrote myself back there again" (2015, 131).

19 Sickels' ambivalence about the rural for himself and his characters is both poignant and productive. While ambivalence is a difficult emotion, it's also an opportunity. Johnson, Gilley, and Gray argue, "[B]eing ambivalent or conflicted is something different from being irrelevant or unimportant" (2016, 16–17).

20 Sickels powerfully addresses the push-pull between urban and rural life in an email exchange wit me:

> I think there is much truth to the notion that queer and trans people live predominately in urban areas—this has been the case for a very long time. In cities, we're free from conservative families and churches, and small-town politics and policing. The anonymity gives us a kind of freedom. And, generally, queers are moving to cities that are more liberal or progressive, and where we can find community. However, not all of us live in cities, and I think economic class certainly complicates this notion. How does a working-class queer person move to New York City or San Francisco? And, it's not just about economics, but about home and comfort.
>
> (Carter Sickels, email message to author, June 21, 2016)

21 Jillian's voyeuristic approach to Dean and Perry reflects a long history of cultural exploitation of Appalachia. As Elizabeth Catte documents, the extensive "visual archive" of Appalachia constructs the land and people as abject victims suffering from regional deprivation and innate biological and cultural inferiority (2018, 58).

22 In an email interview with me, Carter Sickels explains the significance of home in this story:

> I'm … trying to explore the many layers of "home," and how that connects to place and queer identity. Home, for many queers, is fraught with tension, and many of us have a tenuous relationship with the home we come from. … And, yet, [Dean] feels a deep connection with this place; he feels rooted here.
>
> (Carter Sickels, email message to author, June 21, 2016)

However, from the perspective of urban-centric queer theory, Sickels' comments about home, place, and "roots" could be dismissed as sentimental capitulation to neoliberal queer normativity. Dean's fantasy of vacationing in Perry with Jillian is particularly suspect: "It will be Jillian's and my vacation house. She'll work on films, I'll plant a garden, our New Yorker friends will visit for long weekends. We'll have barbecues, sun ourselves at the swimming hole, read under the shade trees" (Sickels 2012, 23). But Sickels complicates this possible reading by using Dean's fantasy to illuminate Jillian's urban-centrism. Her anxious desire to return to the city—"I think it'll feel good getting back to New York. Don't you?" (2012, 41)—conveys her assumption that Perry is not "transgressive" or properly queer enough for her.

23 Sickels' view of queers and nature resonates with sentiments of the "farm boys" in Fellows (1996).

24 Sickels refuses to take an isolationist view of nature, always including vital contemporary environmental and social issues as a way to undercut nostalgia: "When I was a kid, my grandmother and I fished for catfish and trout, but after the coal company started stripping above us, the water turned the color of Tang and most of the fish died" (Sickels 2012, 22).

25 Dean's discovery of his "true self" by way of returning to nature/home place resonates with Parker's analysis of Lili Elbe's statement, "the bog is in me" (Parker, Chapter 1, this volume) . However, it's important to point out that Sickels offers shifting ideas about gender transition and identity in his work. On the one hand, his character, Dean, evokes a model somewhat akin to poststructuralist queer theory:

> People think that the decision to transition is something you've always known, or that one day you experience a single earth-shattering epiphany. Maybe for some it's like that, but for me, for so long, I've both known and not known; I've had experiences that led me here, took me away, and brought me back, a tide I can't predict.
>
> (Sickels 2012, 26)

Here, the continuous changing ocean's tide is a metaphor for transitioning, dovetailing with Dean's stunning "I am a project, I think, that will never be finished" (Sickels 2012, 39). Likewise, in Sickels' personal essay "Bittersweet," he says, "Maybe home is fluid, just like identity, and sexuality and gender" (2015, 79). At the same time, his character Dean uses language that constructs identity as having a stable inner core: "Transitioning is like stepping into another country and yet a country that I already knew from some deep place inside me" (Sickels 2015, 79). Dean describes transitioning as moving "closer to my true self" and says that it means that one becomes "More the person you imagine yourself to be" (Sickels 2012, 38, 75). Because of Sickels' commitment to rural place and home as central threads of identity, I would argue that, overall, he constructs trans identity as a significant thread rather than a core, innate identity. However, I believe it's important to point out the conflicting "core"/fluid models of identity in his work, not because they indicate a monolithic flaw but because they suggest Sickels' interest in delineating the complexities of literary trans geographies. For an excellent analysis of trans literature and identity see Ladin (2011, 2014, 2016) and Michaelson (2012).

26 Sickels' emphasis on the significance of home and rural place, despite the often extreme difficulties, proposes how rural home "dispossesses" queer people—that queer people don't necessarily "disinherit" home, home "disinherits" them. The idea of rural home place as rejecting queers rather than queers rejecting rural place is a subtle distinction, but it's one that challenges the notion of the rural as universally undesirable to queers. As Todd Ruhter—one of Will Fellows' *Farm Boys*—makes clear, "Where I come from is as important as what I am. In fact, it's hard for me to separate the place from the person" (1996, 305).

27 See Peter I-min Huang, "Chinese literature, ecofeminism, and transgender studies" (Chapter 5, this volume).

28 Rural queer studies/trans literature asks us to critically investigate the treatment of the rural in queer studies, culture, and politics. As Colin Johnson suggests, rural queer scholarship inspires us to consider the effect of "rurality's categorical salience in the ongoing work of investigating the history of queer life in the United States" (2013,10). Expanding how we think about the rural also generates a lot of questions about the effects of urban exceptionalism on *urban* queers. In other words, while the field has convincingly made the case that rural queers are casualties of an overweening metronormativity, the scholarship also implies that metronormativity might have deleterious effects on our understanding of *urban* queers. What forms of urban queer life can't we see when the metronormative lens dominates our vision? What is capacious about urban space and who decides? Is there such a thing as metronormative nostalgia, akin to the uncritical hillbilly/country nostalgia? What might an urban-centric focus keep us from knowing about urban complexities and contradictions? For whom is the city a safe haven, and how might this presumption negatively affect urban queers? What roles do gender, race, language, ability, and class play in specific concepts of the urban? How does the city as a "The Promised Land" complicate the lives of urban trans/queer people of color? Are the metaphors of visibility and the closet truly capable of explaining urban queer and trans people? In other words, are cities, to use the words of Lucas Crawford, "primarily, exclusively, or unambiguously liberating for non-normative people" (2016, 131)?

An example of how queer metronormative culture can be constricting, claustrophobic, and small-minded comes from Ash, one of the interviewees in *Trans/Portraits*. Ash explains:

> I'm really just not satisfied with the more common identity descriptors that are used in the Portland, Oregon, queer population. A lot of the queer folks here tend to be really snotty, and if you don't wear enough glitter or have the correct asymmetrical haircut, you're not queer enough to fit in. I'm not really a tomboy because I'm not really into trucks or motorcycles. I'm usually covered in dirt wearing Carhartts. When I think of myself, gender is not the primary identifier I would use. I view myself most strongly in terms of occupation.
>
> (2015, 188–9)

Ash speaks as an *urban* trans farm boy whose experiences and perspectives clash with the The Promised Land image of the city; he points out that the urban queer subject who "counts" as queer embodies a particular physical appearance, displays a keen sense of urban fashion, and emphasizes non-normative identity. The reigning urban social and fashion customs do not speak to his primary identity as a farmer. Although Ash lives in a US city, his approach to trans identity and trans expression more closely resembles that of rural queers, putting him at odds with the queer metronormative world. In a sense, an urban inhabitant such as Ash is a casualty of queer metronormativity.

29 In addition to Sickels' literary work, the idea that we are all potentially attached to inhospitable places emerges in Colin Johnson's book, *Just Queer Folks: Gender and Sexuality in Rural America* (2013)—in particular, chapter 4, "Community Standards: Village Mentality and the Queer Eccentric." In "Community Standards," Johnson discusses contemporary queer activists' and queer scholars' disdain for "benevolent toleration," a phrase that refers to the ways in which rural queers were (and still are) "tolerated" by their towns and communities rather than cherished. Understandably, this is a pattern of treatment that many contemporary queers find unacceptable. However, Johnson wisely suggests that there is a cost to overlooking "toleration" as a category of analysis:

> For one thing, doing so makes it harder to recognize the existence of some forms of queer life, especially those that take shape under circumstances that seem, from our perspective, unlivable on their face. Rejecting social arrangements that depend on tolerance arguably also leads to the denigration of certain forms of queer life because they seem to entail too much negotiation and concession making, as if all forms of life do not involve these things to some extent.
>
> (2013, 109)

There are two profound ideas here. One, as queer scholars, it behoves us to be interested in all forms of queer life—not just those forms that match our ideas of what queer life should look like. Characterizing particular individuals as conformists, sellouts, or closet cases might hinder our ability to consider as many forms of queer life as possible. Second, as Johnson implies, to "some extent" most of us, whether we realize it or not, are complicit in making concessions and negotiations in order to live our lives.

References

Abelson, Miriam J. 2016. "'You Aren't from Around Here': Race, Masculinity, and Rural Transgender Men." *Gender, Place, and Culture* 23(11): 1535–46.

Annes, Alexis and Meredith Redlin. 2012. "Coming Out and Coming Back: Rural Gay Migration and the City." *Journal of Rural Studies* 28: 56–68.

Ash. 2015. "Accidental Activists." In *Trans/Portraits: Voices from Transgender Communities*, edited by Jackson Wright Shultz, 165–95. Hanover: Dartmouth College Press.

Baker, Kelly. 2016. "Out Back Home: An Exploration of LGBT Identities and Community in Rural Nova Scotia, Canada." In *Queering the Countryside: New Frontiers in Rural Queer Studies*, edited by Mary L. Gray, Colin R. Johnson, and Brian J. Gilley, 25–48. New York: New York University Press.

Bauer, Henry. 1996. "Henry Bauer." In *Farm Boys: Lives of Gay Men from the Rural Midwest*, edited by Will Fellows, 59–67. Madison: University of Wisconsin Press.

Brady, Mary Pat. 2016. "The Waiting Arms of Gold Street: Manuel Munoz's *Faith Healer of Olive Avenue* and the Problem of Scaffold Imaginary." In *Queering the Countryside: New Frontiers in Rural Queer Studies*, edited by Mary L. Gray, Colin R. Johnson, and Brian J. Gilley, 109–15. New York: New York University Press.

Butler, Judith. 1990. *Gender Trouble: Feminism and the Subversion of Identity.* New York: Routledge.

Catte, Elizabeth. 2018. *What You Are Getting Wrong About Appalachia.* Cleveland, OH: Belt Publishing.

Crawford, Lucas. 2013. "Transgender Without Organs? Mobilizing a Geo-affective Theory of Gender Modification." In *The Transgender Studies Reader 2*, edited by Susan Stryker and Aren Z. Aizura, 473–82. New York: Routledge.

Crawford, Lucas. 2016. "Snorting the Powder of Life: Transgender Migration in the Land in *Oz*." In *Queering the Countryside: New Frontiers in Rural Queer Studies*, edited by Mary L. Gray, Colin R. Johnson, and Brian J. Gilley, 126–45. New York: New York University Press.

Crawford, Lucas. 2017. "A Good Ol' Country Time: Does Queer Rural Temporality Exist?" *Sexualities* 20(8): 904–20.

Cross, Jim. 1996. "Jim Cross." In *Farm Boys: Lives of Gay Men from the Rural Midwest*, edited by Will Fellows, 76–88. Madison: University of Wisconsin Press.

Doan, Petra L. 2010. "The Tyranny of Gendered Spaces—Reflections from Beyond the Gender Dichotomy." *Gender, Place, and Culture* 17(5): 635–54.

Fellows, Will, ed. 1996. *Farm Boys: Lives of Gay Men from the Rural Midwest.* Madison: University of Wisconsin Press.

Gray, Mary. 2009. *Out in the Country: Youth, Media and Queer Visibility in America.* New York: New York University Press.

Gray, Mary L., Colin R. Johnson, and Brian J. Gilley. 2016. *Queering the Countryside: New Frontiers in Rural Queer Studies*. New York: New York University Press.

Hain, Mark. 2016. "'We are Here for You': The It Gets Better Project, Queering Rural Space, and Cultivating Queer Media Literacy." In *Queering the Countryside: New Frontiers in Rural Queer Studies*, edited by Mary L. Gray, Colin R. Johnson, and Brian J. Gilley, 161–80. New York: New York University Press.

Halberstam, Judith [now known as Jack]. 2005. *In a Queer Place and Time: Transgender Bodies, Subcultural Lives.* New York: New York University Press.

Herring, Scott. 2010. *Another Country: Queer Anti-urbanism.* New York: New York University Press.

Hines, Sally. 2010. "Queerly Situated? Exploring Negotiations of Trans Queer Subjectivities at Work and within Community Spaces in the UK." *Gender, Place, and Culture* 17(5): 597–613.

Holloway, Pippa. 2016. "Manifesto for a Queer South Politics." *PMLA* 131(1): 182–6.

hooks, bell. 2015. "A Constant Mourning." In *Walk Till The Dogs Get Mean: Meditations on the Forbidden from Contemporary Appalachia*, edited by Adrian Blevins and Karen Salyer McElmurray, 11–15. Athens: Ohio University Press.

Howard, John. 1999. *Men Like That: A Southern Queer History*. Chicago, IL: University of Chicago Press.

Howard, John. 2007. "Of Closets and Other Rural Voids." *GLQ: A Journal of Lesbian and Gay Studies*, a Brokeback Mountain Dossier, 13(1): 100–2.

Howard, John. 2016. "Digital Oral History and the Limits of Gay Sex." In *Queering the Countryside: New Frontiers in Rural Queer Studies*, edited by Mary L. Gray, Colin R. Johnson, and Brian J. Gilley, 309–32. New York: New York University Press.

Johnson, Colin R. 2013. *Just Queer Folks: Gender and Sexuality in Rural America.* Philadelphia, PA: Temple University Press.

Johnson, Colin R., Brian J. Gilley, and Mary L. Gray. 2016. Introduction. In *Queering the Countryside: New Frontiers in Rural Queer Studies*, edited by Mary L. Gray, Colin R. Johnson, and Brian J. Gilley, 1–21. New York: New York University Press.

Jones, Tennessee. 2015. "Getting Out: The Grief of Transformation." In *Walk Till the Dogs Get Mean: Meditations on the Forbidden from Contemporary Appalachia*, edited by Adrian Blevins and Karen Salyer McElmurray, 121–34. Athens. Ohio University Press.

Keller, Julie C. 2015. "Rural Queer Theory." In *Feminisms and Ruralities*, edited by Barbara Pini, Berit Brandth, and Jo Little, 155–66. Lanham, MD: Lexington Books.

Kuusisto, Stephen. 2015. Preface to *From Where We Stand: Recovering a Sense of Place*, by Deborah Tall, xi–xiv. Syracuse, NY: Syracuse University Press.
Ladin, Joy. 2011. "Ours for the Making: Trans Lit, Trans Poetics." *Lambda Literary*, December 6. www.lambdaliterary.org/features/oped/12/06/ours-for-the-making-trans-lit-trans-poetics/.
Ladin, Joy. 2014. "I Am Not Not Me: Unmaking and Remaking the Language of the Self." *Lambda Literary*, December 28. www.lambdaliterary.org/features/12/28/unmaking-and-remaking-the-language-of-the-self/.
Ladin, Joy. 2016. "'Split It Open and Count the Seeds': Trans Identity, Trans Poetics, and Oliver Bendorf's *The Spectral Wilderness*." *TSQ: Transgender Studies Quarterly* 3(3–4): 637–48.
Lee, Ang. 2005. *Brokeback Mountain*. DVD. Focus Features, USA.
Michaelson, Jay. 2012. Book review of *Through the Door of Life*. "Through the Door of Life: A Jewish Journey between Genders. *Lambda Literary*, May 16. www.lambdaliterary.org/reviews/05/16/through-the-door-of-life-a-jewish-journey-between-genders-by-joyladin/?hilite=%22Jay%22%2C%22Michaelson%22.
Olivia. 2015. "You Can't Do It Alone." In *Trans/Portraits: Voices from Transgender Communities*, edited by Jackson Wright Shultz, 77–99. Hanover: Dartmouth College Press.
Range, Melissa. 2015. "Outsider Appalachian." In *Walk Till the Dogs Get Mean: Meditations on the Forbidden from Contemporary Appalachia*, edited by Adrian Blevins and Karen Salyer McElmurray, 110–20. Athens: Ohio University Press.
Ross, Marlon B. 2005. "Beyond the Closet as Raceless Paradigm." In *Black Queer Studies: A Critical Anthology*, edited by E. Patrick Johnson and Mae G. Henderson, 161–89. Durham, NC: Duke University Press.
Ruther, Todd. 1996. "Todd Ruther." In *Farm Boys: Lives of Gay Men from the Rural Midwest*, edited by Will Fellows, 304–10. Madison: University of Wisconsin Press.
Schweighofer, Katherine. 2016. "Rethinking the Closet: Queer Life in Rural Geographies." In *Queering the Countryside: New Frontiers in Rural Queer Studies*, edited by Mary L. Gray, Colin R. Johnson, and Brian J. Gilley, 223–43. New York: New York University Press.
Shultz, Jackson Wright, ed. 2015. *Trans/Portraits: Voices from Transgender Communities*. Hanover: Dartmouth College Press.
Sickels, Carter. 2012a. *The Evening Hour*. New York: Bloomsbury.
Sickels, Carter. 2012b."Saving." In *The Collection: Short Fiction from the Transgender Vanguard*, edited by Tom Leger and Riley MacLeod, 21–52. New York: Topside Press.
Sickels, Carter. 2015. "Bittersweet: On Transitioning and Finding Home." In *Walk Till the Dogs Get Mean: Meditations on the Forbidden from Contemporary Appalachia*, edited by Adrian Blevins and Karen Salyer McElmurray, 73–9. Athens: Ohio University Press.
Smith, Aaron. 2015. "For Better or Worse, This was My Country." In *Walk Till the Dogs Get Mean: Meditations on the Forbidden from Contemporary Appalachia*, edited by Adrian Blevins and Karen Salyer McElmurray, 194–7. Athens: Ohio University Press.
Soderling, Stina. 2016. "Queer Rurality and the Materiality of Time." In *Queering the Countryside: New Frontiers in Rural Queer Studies*, edited by Mary L. Gray, Colin R. Johnson, and Brian J. Gilley, 333–48. New York: New York University Press.
Weston, Kath. 1995. "Get Thee to a Big City: Sexual Imaginary and the Great Gay Migration." *GLQ: A Journal of Lesbian and Gay Studies* 2(3): 253–78.
Williams, Tennessee. 2004. *Cat on a Hot Tin Roof*, revised edition. New York: New Directions.

8 Sexuate ecologies and the landmarking of transgender cultural heritage in Australian schools

Nicole Anae

Introduction

In 2003, UNESCO adopted the following convention safeguarding and defining "cultural heritage as the practices, representations, expressions, knowledge, skills … as well as the instruments, objects, artefacts and cultural spaces associated therewith–that communities, groups and, in some cases, individuals, recognize as part of their cultural heritage" (UNESCO 2003). The convention makes a distinction between "tangible" and "intangible" heritage in an effort to include oral stories, traditions, histories, languages and expressions, the creative and performing arts, rites, festivals and rituals, and traditional ecological practices and knowledges, as well as traditional craftwork and decorative arts (Liu 2011, 316). Kearney argues that in the question of tangible and intangible perceptions with respect to heritage, "the only imperative status of tangible is held by the human actor and agent, as physical embodiment of culture and heritage" (2009, 211).

It is clear that in exploring interpretations of "cultural heritage" there is in the very term an implied audience, and that heritage sites perform a direct or indirect communicative function in a narrative sense. From the standpoint of narrative theory, for instance, heritage sites act as interlocutors; that is, at best we can say these sites function as "the narrator narrates, not *to* a narratee but is overheard by a (to the narrator) non-existent listener" (Behrendt and Hansen 2011, 237). It is for this reason that Stuart Hall's suggestion to define heritage as a "discursive practice" is particularly salient (2005, 25). Considered as a form of narrative, cultural heritage sites construct and perpetuate a sense of belonging to a nation. "Heritage" too, therefore, simultaneously interrogates the Other and its relation to the discursive framework upon which "relation to Nation" is constructed. Here, in determining, protecting, and promoting sites of transgender heritage, the narrative practices that institutions and groups use to posit ecological meaning through bodies—"sexuate difference" (Irigaray 2008, 77) and the subjectivity of Otherness—become crucial to according and defining spaces of cultural *heritage interest.*

Cultural heritage is assessed as an invaluable pragmatic movement at a time when the interrogation of bodies within and against environmental and architectural boundaries is more important than ever. In Australia, museums such as

Museum Victoria and Museums Australia have revised their curatorial and collecting policies to ensure that LGBTQ material is actively collected, with heritage institutions such as the State Library of South Australia and the Victorian Archives Centre, Melbourne, among others, holding in-house exhibitions focusing specifically on LGBTQ histories and material culture. Organizations such as Australian Lesbian and Gay Archives (ALGA), established in 1978, together with the Women's Liberation and Lesbian Feminist Archives, in Melbourne, and the Pride History Group, in Sydney (Davison 2011), among others, have also been collecting and documenting the material culture of LGBT people and communities for decades, and in so doing actively address LGBTQ histories and audiences. Coalitions such as the Australian Gay and Lesbian Archives in Melbourne, Sistergirls and Brotherboys (gender-diverse Aboriginal Australians from around the country), and landmarks such as the Rainbow Walk in Adelaide's Light Square, Sydney's Gay and Lesbian Holocaust Memorial, and Yarra Council's statue commemorating the courage of the local LGBTQ community, among many others, achieve important outcomes in connecting lesbian, gay, bisexual, transgender, intersex, and queer identity to place. The preservation of sites of historic and cultural importance to LGBTQ communities elevates the concept of landmarking to an issue of human rights as much as informing a broader cultural push in acknowledging transgender individuals within the discourse of nation and belonging: "There is a class of landmarking that imbues whatever structure is there with the power of its history" (Byard, quoted in Zara 2015).

Approaches to historic preservation, recognition, and landmarking are clearly at play in cultural heritage practices and the perception of transgender ecologies as an issue of heritage interest. This chapter approaches the concept of "landmarking" from Michael Sorkin's perspective as "a very frail bulwark, finally answerable only to staid historical routines: unfortunately you can't landmark people's lives" (1991, 364). But there are opportunities to explore the discursive practices and cultural heritage practices that "landmark" transgender lives within school spaces as much as other built and non-built environments. English Heritage's (2008) position regarding "social value"—as being associated with "places that people perceive as a source of identity, distinctiveness, social interaction and coherence" (32)—is an important concept in theorizing how school ecologies ascribe sexuate difference to spatial and natural structures in according and preserving cultural heritage.

It is important to note at this point that this examination uses specific terminology that conventionally resists consensus as definable concepts. Mathy, for instance, observes that "In the 1990s, transgender identity emerged as a concept to define individuals who live full- or part-time as members of a sex different than their sex of birth, frequently with hormonal support but without pursuit of (or desire for) surgical reassignment" (2003, 327). Speer, on the other hand, distinguishes between the terms "transsexual" and "transgender" and argues that "transgender is often used in a political context by transgender activities in order to avoid medical categorization" (2010, 154). In this chapter, we will use the word "transgender" to describe persons identifying as a gender that does not correspond to their physiological, anatomical, and/or genetic makeup. In this respect, "transgender" is more than behavioral (Hill 2006, 3). While this term may or may not precisely cover all members of the transgender community, we use it here in an effort to circumvent a more complex theoretical examination of sexual orientation and gender identity.

Transgender cultural heritage

According to Rebecca Dierschow, the "lesbian, gay, bisexual, and transgender (LGBT) community is one example of a community that has been largely invisible throughout history yet has created significance and distinctive traces and spaces" (2016, 95). Yet it is clear that in exploring interpretations of "cultural heritage" there is the recognition that specific spaces encompass what John Schofield characterizes as "iconic places of deeper history, places associated with notable events and individuals who did more than most to create the contemporary world" (2016, 3). These spaces speak to an implied audience; performing a direct or indirect communicative function in a narrative sense: a "discursive practice" (Hall 2005, 25). Considered as a form of narrative, cultural heritage sites construct and perpetuate a sense of belonging well beyond place. Here, in identifying and acknowledging sites of transgender heritage, the narrative practices that institutions and groups use to posit ecological meaning through bodies—"sexuate difference" (Irigaray 2008, 77)—become crucial to a globalized enterprise of cultural and *human heritage*.

For Irigaray, the "sexuate" is not analogous with "sexual," with the latter equated by Irigaray as the individual's sexual "object" choice. The concept of "belonging" in Irigarayan thinking seems to mediate the gap between gender dualities:

> *I* is never simply mine in that it belongs to gender. Therefore, I am not the whole: I am man or woman. And I am not a simple subject, I belong to a gender. I am objectively limited by this belonging … I belong to a gender, which means to a sexuate universal and to a relation between two universals.
> (Irigaray 2004, 10)

Even though Irigaray resists the expansion of her work on sexual difference to transgender and transsexual persons, more and more scholars are demonstrating that Irigaray's concept of sexual difference can, in fact, be applied to transgender and transsexual narratives (Murphy 2007; Salamon 2010; Poe 2011). Gayle Salamon addressed the material body in sexuate terms to expand the concepts of "what 'counts' as a body" (2010, 7). "Indeed," claims Murphy, "the rift between sex and gender is implicit on the logic of transgenderism and transsexuality, as the figure of one's gendered imaginary is not always commensurate with the contours of one's material body" (2007, 89). Danielle Poe's interpretation of first-person narratives of transgender and transsexual persons argues that these individuals can be accommodated within Irigaray's notion of sexual difference (2011, 126). Poe turns Irigarayan thinking about sexual difference back on itself "to provide a theoretical groundwork for interpreting the narratives of transsexual people who describe their experience of crossing genders as visibly becoming the people that they always knew themselves to be" (122).

In this analysis of landmarking transgender cultural heritage, I aim to incorporate the Irigarayan notion of "belonging" with Gail Schwab's exploration of inter-sexuate and inter-subjectivities in the classroom (2016). While Schwab's primary interest lies in gender differences in language, her interest in language use in the classroom draws parallels between Hall's concept of "discursive practice" and the ways in which the ecologies of schools can be regarded as sites of identity-making, what Schwab terms "inter-sexuate inter-subjectivity" (149). Beyond the confines of the classroom landscape, the interest of this chapter is to explore how transgender students might use language differently and how this difference might "create inter-subjectivity between subjects in sexuate difference, not just to study it, not just to confirm is absence from language, but to make it happen" (147). Here, we argue that schools also engage in landmarking transgender heritage, and that these ecologies *are* authored by those staking a claim for the school site as one of ecological significance.

Schools and the transsexual body: an ecological question

More and more research is emerging not only expounding the importance of incorporating LGBTQ themes and histories into the school curriculum (Abreu and Fedewa 2016; Chappell, Ketchum and Richardson 2018), but also interrogating the challenges and difficulties in promoting such integration (Boske 2015). However, relatively little scholarly work exists theorizing transecological practices of LGBTQ heritage preservation and landmarking within schools. Caroline Wilson (2015) brings the philosophical question of sexual difference between classroom teachers, female and male, directly into the school's educational landscape, drawing specifically upon Irigarayan thinking, claiming:

> If the idea is that change can come about through women authorising one another in their experience as women, as sexuate subjects (and the same applies to the male sexuate difference), bringing change into the classroom would require an *a priori* recognition of this fact. It might lead to a situation in which men and women teachers would understand that embracing this sexuate difference and working *with* it could generate a new awareness of how relationships between them might change.
>
> (34, emphasis in original)

For the purposes of this chapter, theorizing the school ecology as transgender heritage site, it is possible to adopt Wilson's reading of "sexuate subjects" to school students as much as to female and male teachers. Put another way, students also authorize one another within the school ecology, and that through "embracing this sexuate difference and working *with* it" there is the potential for raising student awareness of the possibilities for relational change both within and beyond that ecology. From this perspective, "Human existence is thereby seen as both constitutively sexuate, and constitutively relational" (Jones 2015, 11).

Wallace Stegner once claimed that "whatever landscape a child is exposed to early on, that will be the sort of gauze through which he or she will see all of the world afterward" (Nabham and Trimble 1994, 121). The quote offers rich potential in suggesting transecological associations between childhood and nature—the landscape of childhood, the geography of memory, school as landscape, and the storying of "scapes": landscapes, school-scapes, body-scapes. This interconnection between concepts of the internal and external landscape is a nexus that Cobb (1977) has described thus: "At the level of participation in nature during childhood, there is fusion between emotion as the energy of spirit and the spirit of place as the energy of the behaving world" (32).

The emergent albeit less written-about discourse is one that explores the transgender student and concepts of "school ecology." "School ecology" is a term embracing the various elements that define the school. These include physical characteristics, such as architecture, statistical characteristics, such as the socio-economic profile of the school's catchment area as well as total student enrolments, functional characteristics, such as school policies regulating student behavior and expectations, as well as affective characteristics, such as school–staff and school–student interrelationships (Waters, Cross, Shaw et al. 2010, 384). The nature of schools, their ecologies, and their primacy as sites for constructing realities, are equally "the spirit of place as the energy of the behaving world." That the importance of school ecology to student learning success has been well documented (Ristuccia 2013, 255) therefore suggests that school plays a significant role in the politics of identity as much as school and community belonging. Just as the UNESCO convention distinguishes between "tangible" and "intangible" heritage, similar distinctions can be made with reference to "intangible" aspects of school ecology. These "intangibles" include: the extent to which school members are engaged in its community in meaningful and influential ways; the interpersonal and emotional relationships that show the school community as caring and supportive; as well as well-being and zero-tolerance initiatives, among other things.

Theorizing about the intersection of transgender issues and ecology through the lens of cultural heritage is becoming especially urgent given the recent pressures on *school ecology* consequential to political reform and transgender policy debates. In the current milieu, not only do these pressures indicate the infestation of the worlds of principals and teachers (Sergiovanni 2000), but also of the problematizing of transgender people and the child's body within school ecologies. Just as decision-making represents the "lifeworld" and "lifeblood" of teachers and principals (Sergiovanni 2000, 2005), so too do matters of national policy hold serious long-term consequences for transgender children in school, and by extension, the school ecologies which are shaped and manipulated as a result. By adopting a definition of the school environment encompassing not only natural resources but also "the characteristic aspects of landscape and property forming part of the cultural heritage" (Brans 2001, 10) the crossings between the transsexual body of the child as both "objects of human origin" and an ecology upon which place is given significance as sites of contention in the politics of identity and heritage.

In May 2016, US President Barack Obama defended his administration's efforts to compel schools to permit transgender students to use their preferred bathroom under the premise that the United States was required to ensure dignity and respect for all schoolchildren. In his response to BuzzFeed, President Obama reiterated that his administration's policy on transgender bathroom use would offer protection to young people already susceptible to bullying and harassment (Geidner 2016). Mr. Obama told the website: "Kids who are sometimes in the minority, kids who have a different sexual orientation, who are transgender are subject to a lot of bullying potentially. They are vulnerable" (Geidner 2016). While the Obama administration could only make the policy, non-compliant school districts could potentially face either lawsuits from the federal government and/or a loss of federal funding. In Massachusetts, a high school in Ipswich opened an all-gender bathroom so that cisgender students could, as the Obama administration directive allows, use the bathroom that corresponds to their preferred gender identity. According to Jeff Perrotti, Director of the Safe School Program for LGBTQ Students, a joint initiative between the Massachusetts Commission on LGBTQ Youth and the Massachusetts Department of Elementary and Secondary Education, "An all-gender bathroom allows students an option who don't feel comfortable in a gendered bathroom.... There's a recognition we may need more options" (Mac Alpine 2016). However, in 2017, as noted by Anna Bedford in her Introduction to this collection, "a month after taking office, President Trump rolled back protections for trans students that had ensured their right to use school bathrooms that matched their gender identity" (p. 2, this volume).

According to Peters, Becker, and Davis' piece for the *New York Times*, when President Trump "rescinded protections for transgender students that had allowed them to use bathrooms corresponding with their gender identity," he effectively overruled "his own education secretary and plac[ed] his administration firmly in the middle of the culture wars that many Republicans have tried to leave behind" (Peters et al. 2017). However, Trump's backflip had little legal impact given reports that "The same federal and state laws that protect transgender people against discrimination are still in place, and Trump's move has no effect on them" (Bendery and Farias 2017). Coincidentally, in Australia around the same time, the Education Department implemented in South Australian public schools a new policy for transgender and intersex students providing for the use of facilities corresponding to the gender with which they identified (Department for Education and Child Development 2016, 7).

According to statistics produced by Trans Student Educational Resources (2016), almost 80 percent of transgender students feel unsafe at school, with nearly 60 percent stating they have been bullied and harassed at school within the past year, compared with around 30 percent of their peers. Also, half of the students surveyed claimed they had been physically assaulted, with one in five having been forced out of their homes. Similarly, according to mental health and well-being research focusing on gender-diverse and transgender young people conducted in Australia in 2014 (Smith et al. 2014, 12):

> Participants who did not feel supported by their teachers were over four times more likely to leave school if they experienced discrimination than those with teacher support. Inclusive schools were those where leadership and teachers tried to address students with their preferred pronouns, were flexible about uniform and toilet arrangements, took a stand against bullying, and aimed to be accommodating to the individual's needs. Improvements to sexuality and puberty education, which are inclusive of these students, are needed.

The inclusion of statistical data focusing on health, well-being, and learning outcomes for LGBTQ students in a discussion of the landmarking of transgender cultural heritage aims to recognize: (1) the growing significance school ecologies play in the social recognition of transgendered young people, and (2) the significant role discursive practices play in the emerging visibility of LGBTQ children and young people and understandings of transgender identity.

The Australian Safe Schools Program: a case study of transgender cultural heritage

The Australian government has responded to increasing numbers of transgender children in educational institutions with the "Safe Schools Program." While the program's political support, as much as its coverage in the Australian media, is characteristically polarized—denounced by some as a "sexual indoctrination program" producing "an epidemic of transgender children" as its legacy (Devine 2017), and others as having "the potential to save lives" (Tomazin 2016)—this program is currently supported by "mental health organisations *beyondblue* and *headspace*, as well as the Australian Secondary Principals Association and the Foundation for Young Australians" (Brown 2016). This government-funded program was launched on June 13, 2014 (Middleton 2016) and aims to foster safe and inclusive school environments for transgender, intersex, gender-diverse, and same-sex-attracted students.

> Safe Schools Coalition Australia has assisted hundreds of schools across the country in their efforts to actively promote safety and inclusion for the benefit of the whole school community. This includes working in partnership with government and independent schools, schools in diverse geographic locations, and faith-based schools.
>
> (Foundation for Young Australians 2016)

While this program is controversial, with commentators such as Dr. Lucy Nicholas describing Members of Parliament critical of the Safe School Program as “white, cisgender, heterosexual male politicians” (Nicholas 2016; Donnelly 2016), what is particularly interesting about it from the cultural-heritage studies perspective is the way it accommodates a definition of the school as generating connections between the ecology and cultural heritage, while simultaneously functioning as an ideological landscape in which the child generates and sustains individual identity as a form of cultural heritage.

“Briella day”

Exploring schools as a way to generate connections between ecology and cultural heritage offers great potential in examining this institution’s simultaneous functioning as an ideological landscape in which the child sustains and generates individual identity as a form of cultural heritage. In the case of 6-year-old transgender child Briella Carmichael, for instance, her school, Cranbourne South Primary (Victoria), part of the “Safe Schools” network, held a “Briella Day” during which Briella received a new reader and book bag stating her chosen name as “Briella” (Bailey 2016). In this sense, the school is analogous to “social ecology” if “social ecology” is “concerned with the relationship between people [school students] and their natural and social environment” (Rummler 2014, 3). This child was born “male” (and named “Baylin”) but identifies as female. Here, the observance of such moments becomes part of the “school ecology”—that is, contributing to both the tangible and intangible elements defining the school’s culture.

While for Zandvliet the phrase “Ecology of School” is an umbrella term for “the development and inception of a variety of unique learning environments and [...] dynamic interactions between people, places and curriculum” (2013, viii), the definition of school ecology extends well beyond the rooms in which learning and teaching takes place. These elements include physical characteristics, such as architecture and the design of buildings, statistical characteristics, such as the socio-economic profile of its catchment area, total student enrolments, functional characteristics, such as school policies that regulate student behavior and expectations, and affective characteristics, such as staff and student interrelationships (Waters, Cross, Shaw et al. 2010, 384).

Eisner’s (1998) five dimensions of educational settings—namely intentional, curricular, pedagogical, structural, and evaluative—can assist in understanding sites within the school ecology and the discursive practices taking place within them:

- The *intentional* dimension refers to explicit and implicit but activated aims or goals of the school/curricula.
- The *curricular* dimension refers to how educators put their ideas, aims, or goals into action: what is being taught and how it is being taught.
- The *pedagogical* dimension refers to how educators operationalize the curriculum; that is, how individual teachers facilitate the curriculum and bring

into it their own personality, passions, strengths, weaknesses, and intentions (Eisner 1998).

- The *structural* dimension concerns "how the organizational envelopes we have designed affect how education occurs" (Eisner 1998, 75). This dimension includes elements such as time management, use and availability of space, exterior needs, obstacles, and support.
- The *evaluative* dimension "concerns the making of value judgments about some object, situation, or process" (Eisner 1998, 80), including the assessment and evaluation of student work and performance.

Represented graphically in terms of landmarking transgender heritage within schools, the interrelationships between these ecological sites might look something like Figure 8.1.

In this sense, "Briella Day" cuts across each of Eisner's five dimensions of educational settings: intentional, curricular, pedagogical, structural, and evaluative.

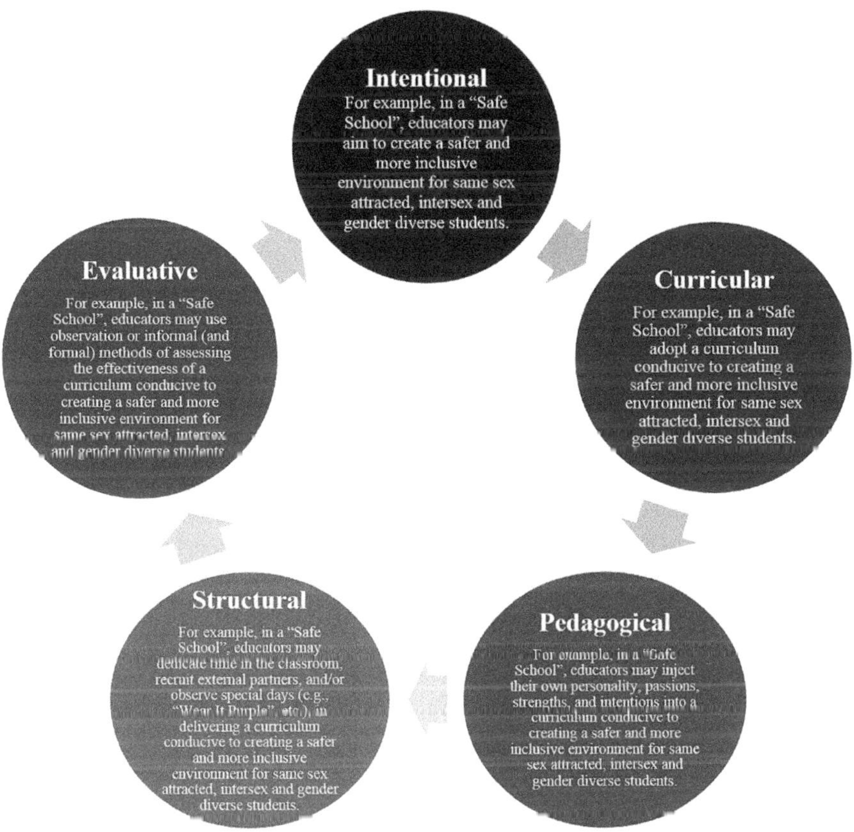

Figure 8.1 Eisner's five dimensions and the school ecology as transgender heritage site (adapted from Eisner 1998).[1]

Beyond simply a "day," "Briella Day" holds implications for how these four types of landmarking the school's ecology generate meaning within and beyond the school environment. As Schlamb asserts, the process of creating landmarking significance "evolve[s] from a place of telling to a place of knowing" (2017, 48). "Briella Day" thus articulates a "discursive practice," to use Stuart Hall's term, and as such landmarks a form of LGBTQ cultural heritage. In receiving a new reader and book bag labeled "Briella," the narrative practices that posit ecological meaning through the transgendered child's body—"sexuate difference" (Irigaray 2008, 77)—also evolve a metaphorical understanding of Briella's subjectivity "from a place of telling to a place of knowing" (Schlamb 2017, 48) in an educational enterprise of cultural and *human heritage*. Cranbourne South Primary (Victoria) both marks and celebrates its otherness as a ("Safe") school ecology as much as a site of transgender cultural heritage.

Safe Schools advocate a policy of "sexual openness" (Wade 2016) which enables, and perhaps also encourages, transgender children to self-authorize. The context in which such self-authorizing takes place is one designed to educate children and young people about sexual and gender diversity.

> All the kids were really excited for her [Briella] to come to school and Safe Schools said "what would you do if you had seen Briella sitting by herself or feeling sad?" and the kids were like "I will go up and cuddle her and say she can play with me" like it was honestly, I had to hold back the tears, it was so sweet.
>
> (Kirra Carmichael, quoted in Sunday Night 2016)

If the idea is that change can come about through children authorizing one another in their experience as children, as *sexuate subjects*, bringing systematic change into aspects of the school ecology from an educational standpoint would require both a theoretical and observational or experiential recognition of this fact.

An Australian website called Minus18, which attracts partial funding from the state government, also provides students with information on how to bind their breasts and "tuck in" male genitalia. Minus18 is promoted by the Safe Schools Coalition and offers readily available instructions on how to deal with "chest dysphoria" and includes a total of seven alternative binding methods, and safety tips to alleviate issues related to body dysphoria (McLeish 2014). Minus18 website states that this platform "is Australia's largest youth-led network for gay, bi, lesbian and trans teens" (Minus18).

The mobilization of programs, such as Minus18 and Safe Schools, among others, stresses the idea that transgender heritage within school ecology is fundamentally a "discursive practice" in the sense defined by Stuart Hall (2005), one which generates a sense of belonging to place and community. What are LGBTQ events such as Rainbow Day and Purple Day if not modes of discursive practices connecting transgender cultural heritage to school ecologies as domains of place and belonging? Contemporary and emergent theories of transgender cultural heritage must therefore recognize the growing significance school ecologies play in both the social recognition of transgendered young people and emerging understandings of the discursive practices shaping that ecology with respect to transgender identity.

Educational leaders conventionally promote and proclaim their affiliation with Safe School's ethos using specific discursive practices linking the school's ecology to aspects of systematic change from a heritage standpoint. According to Safe School proponent and principal of Melbourne's Overnewton Anglican Community College, Jim Laussen (2016), for instance,

> We have never been pressured to promote homosexuality as a preferred lifestyle, to encourage students to come out, to teach children about homosexual acts, to teach children how to bind their chests. Instead, the Safe Schools Coalition has helped us teach our students how to better navigate the differences that they see each day.

In fact, Laussen observes the impact of the Safe School Program's discursive power as a cultural change agent in the claim: "*When our senior footballers tell their opponents not to use 'gay' as a derogatory term, we know that the Safe Schools program is having an impact*" (2016, emphasis in original).

"Wear It Purple Day"; August 31, 2018

Wear It Purple is a student-led, not-for-profit organization dedicated to assisting GLBTIQ youth at risk. "Wear It Purple Day" is its annual event (Figure 8.2). The organization is run by and for students, with its central aim being to put an end to ignorance, bullying, and youth suicide.

According to the Wear It Purple website:

> On September 22nd 2010, 18 year old Tyler Clementi threw himself off the George Washington Bridge in New Jersey. He had just been publically [*sic*] "outed" as gay by his roommate, who video streamed footage of his sexual encounter with another man on the internet without his knowledge or permission. A media frenzy began, and report after report poured in about the individual stories of young people who were committing suicide because of bullying and homophobia.
>
> (New South Wales Teachers Federation 2012)

Wear It Purple Day, like "Briella Day," similarly cuts across each of Schlamb's (2017) nexus identifying four types of landmarks natural and built, historical, collective, and human—with powerful meanings and significance regarding the landmarking of transgender lives extending well beyond simply a "day," and well beyond the school ecology itself.

Schlamb's (2017) four types of landmarks also proves useful in understanding the sites comprising the school ecology and the "discursive practices" taking place within each:

Figure 8.2 "Wear It Purple."

- Natural and built landmarks refers to the physical elements perceived through the five senses.
- Historical landmarks refers to both the literal and figurative signs of the past.
- Collective landmarks refers to the built objects encouraging community value.
- Human landmarks refers to the interpersonal and corporal exchanges with others that inspire landmarking.

Represented graphically in terms of landmarking transgender heritage within schools, the interrelationships between these ecological sites might look something like Figure 8.3.

According to Australian Human Rights Commissioner Tim Wilson (2014–2016):

> Wearing purple is a way for some students to send a message to gay and lesbian students that school is a safe environment for them whether they're open about their sexuality, or not. Is it political? Only to the extent that sending a message that schools should be a bully and harassment-free zone is.

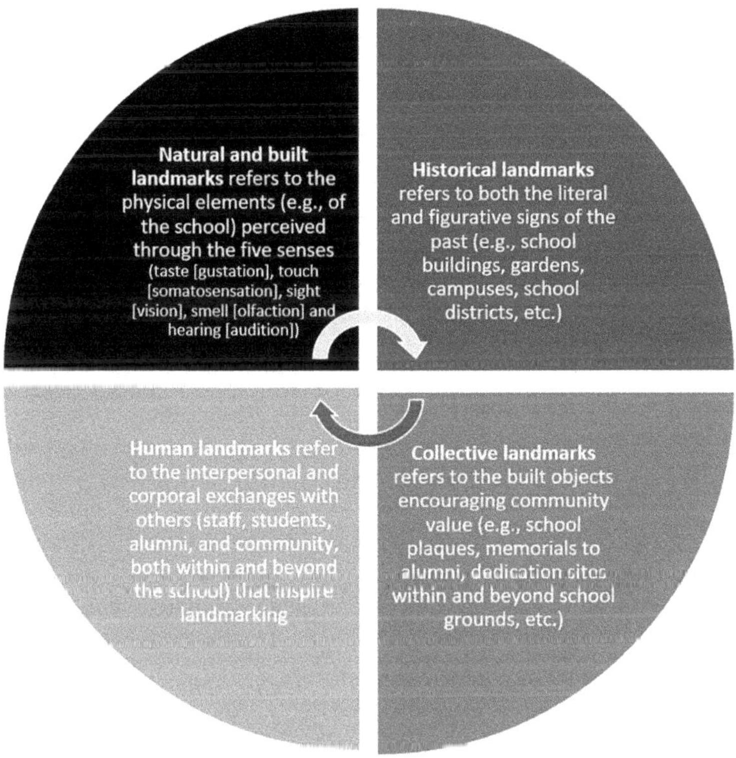

Figure 8.3 Schlamb's four types of landmarking and the school ecology as transgender heritage site (adapted from Schlamb 2016).[2]

> The purpose is to make sure schools can be a safe place for everyone to learn and develop their full potential—regardless of who they are.
>
> (Wilson 2015)

This quote is especially significant in illustrating how political leaders also conventionally promote and proclaim their affiliation with the Safe School's ethos using specific discursive practices linking the school's ecology to aspects of systematic change from a heritage standpoint. Former Australian Human Rights Commissioner Tim Wilson's discursive practice focusing on Wear It Purple, for instance, as a pro-diversity and inclusivity statement, reflects broader socio-cultural developments in transgendered schoolchildren's relationship with the environment, such as the issue of transgendered children competing in track and cross-country races. There have been instances in Australia where children born male and identifying as female have competed as female in these events (Kohlbacher 2015). When a transgender child born male decided to compete in a Perth primary school cross-country race as female, parents were informed in a letter from the school principal that the student would complete as a girl (Brodal 2015). In this example too, the school principal's letter represents a "discursive practice," and as such a form of cultural heritage. In publicly disseminating an acknowledgment of the child's participation in this athletic event, the narrative practice both literally posits ecological meaning through the transgendered child's body—"sexuate difference" (Irigaray 2008, 77)—while simultaneously functioning as a metaphor that defines the subjectivity of Otherness in an educational enterprise of cultural and *human heritage*.

In Western Australia, the Equal Opportunities Act of 1984 protects the rights of children to identify however they choose and also preserves the child's rights to be protected "from discrimination in terms of which uniform she wore, what sport teams she played on and what bathrooms she used, regardless of whether it made anyone uncomfortable" (Young 2015). However, the incident exposed ruptures in the school community given that the child placed fifth in the race, and therefore qualified to represent the school at a regional meet. A parent complained to their local Member of Parliament Peter Abetz (Southern River MLA), who argued that the child had a biological advantage over other competitors in the race. He claimed: "What would have happened if he came first in the race? What would that have created? If I was a girl I would have said that was unfair. The thing is, the child is biologically a boy. That's what he is. It's a scientific fact" (Roberts 2015). Here, in Abetz's rhetoric, the transgender child's relationship to their ecology is somehow distinct from other children, and not only distinct but given a specific context (the athletic cross-country event) that stages those distinctions in a plain and definite manner.

Implicit within Abetz's statement is the view that transgender children not only exert greater control over their physical bodies in the context of a sporting event but that, by extension, they also experience and affect their natural environment differently. Take, for example, this claim by 16-year-old "Bailey," born female but identifying as male:

> I had to drop one of my favourite subjects, which is sport because I can't handle my embarrassment having to change in the female change room. I would really love to get involved in many activities that I can represent my school like inter-school sport competition but I am too embarrassed as I'll be force to play in the girl team, or if I'm not representing my school for sport competition, then other competitions will require myself to dress in full school uniform, which I'll have to wear the girl pants and I hate it.
>
> (Bailey, 16 years old, quoted in Smith et al. 2014, 58)

Bailey's discourse supports the argument that just as schools engage in landmarking transgender heritage, these ecologies *are* authored by those staking a claim for the school site as one of ecological significance. Bailey's disclosure both generates and speaks to the kinds of "discursive practices" used in the context of school sporting events as a form of cultural heritage. This view is in fact echoed in the reflections of transgender cross-country runner Ben Christianson, who participated in both boys' and girls' cross-country teams at Cedar Falls High School in Cedar Falls, Iowa. The then 18-year-old Christianson, born female but identifying as male, claimed in an interview, "I love the physical aspect of it [cross-country running]. Cross-country is such a beautiful sport. Like, you're running on the grass, all the courses are really beautiful. [You know] it's a sport where you're basically closest to nature. And I love being outside" (Zamora 2016).

Christianson's interview offers a fascinating narrative that maps the relationships between school ecologies and school-based practices of cultural heritage. Christianson talks about competing with the female cross-country team and his discomfort with both the sporting apparel he wore at public events (e.g., "women's team" printed on his event clothing), and also the gendered cheers (e.g., "Go ladies") of race-watchers on race day. What Christianson speaks to are the "discursive practices" used in the context of his cross-country race meets as a form of cultural heritage.

These are narrative practices that literally posit ecological meaning through sexuate difference while concomitantly defining the subjectivity of Otherness in an educational enterprise of cultural and *human heritage*. The example illustrates the issue of binary affirmation and pronominal usage within the school setting. Terms such as "she" and "he," for instance, "fix" gender along a binary spectrum of either male or female. Gender neutral terms are gaining acceptance. These include "ze," "hir," or "zir … a *gender-neutral pronoun* popularly employed by queer and *transgendered* persons to replace the gendered pronouns 'his' and 'her' " (Drouin 2014, 233), among other possibilities.

"About a girl": the story of Georgie Stone

The significance of school ecology in the heritage practices of young transgender students is vividly illustrated in the story of Georgie Stone. In 2010, at the age of 11, Georgie, born male but identifying as female, became the youngest person in Australia to be granted pubertal suppression by the Family Court of Australia. *Pubertal suppression* is a process also known as "hormone suppression therapy." Hormone blockers taken during pre/early pubescence "block the hormones associated with one's natal sex and prevent/pause development of gender markers that are incongruent with one's gender identity" (Tando 2016, 144). Strickland claims that Australia is the only country in the world requiring transgender children to petition the court for this kind of treatment (2014). Georgie Stone's case set something of a precedent in Australian family law. The Family Court of Australia determined in 2013 that children were no longer required to petition to court for stage one treatment.

When, as a consequence to Georgie's story, the then Premier of Victoria, Daniel Andrews, visited her school in 2016, the coalescence between the role of the school in landmarking this event contributed significantly in altering the school's own ecology. According to Georgie:

> Daniel Andrews came to my school to meet me, the principal, my mum, my brother, a few of my friends as well to talk about the importance of the Safe Schools Coalition. The safe schools coalition goes to schools and helps both the students and the teachers understand the importance of a safe environment for LGBTIQ kids. I told Daniel Andrews about my experiences of being bullied. He asked my friends about what it was like when I came out to them.
>
> (ABC 2016)

In this instance too, the school generates intersections between its ecology and cultural heritage. The instance of a high-profile political figure visiting a school for the express purpose of acknowledging the social campaigning of a transgender student offers a rich portrait of the school simultaneously functioning as an ideological landscape in which the child sustains and proactively generates individual identity as a form of cultural heritage.

While Georgie Stone did help change Australian law, the victory was only partial in that it did not abolish the mandate for the court-sanctioned authority to undertake stage two puberty-blockers: irreversible gender-affirming hormones (Cohen 2016). Establishing that the teenager is capable of informed consent is known as the "Gillick competent."

"Being Me": the story of Isabelle Langley

Twelve-year-old Isabelle Langley was born male but identities as female. She "came out" to her parents at the age of 11. She also disclosed to her school principal and even wrote a letter to her classmates explaining her need to be identified as female, which was shared publicly at a school assembly. Australian *Four Corners* featured Isabelle's story in a program entitled "Being Me" in 2014. The feature explained Isabelle's story in both human as well as statistical terms, including studies showing the high suicide and self-harm rates among young trans-individuals not receiving treatment. Isabelle's pediatrician, Dr. Michelle Telfer, at Royal Melbourne Hospital's gender clinic, claimed that referrals to the clinic numbered only one case in 2003, while the figure reached 100 in 2014 (Cohen and Scott 2014). In February 2016, Isabelle, together with her sister, Hattie, who was 9 years old, and her parents Naomi Langley and Andrew McNamara, traveled to Australia's capital Canberra to meet with Members of Parliament. Given that Isabelle is now going through male puberty, she is seeking access to stage two cross-sex hormone treatments within the next two to four years in order to permanently affirm her gender as female. For Isabelle, "I am a girl, I was born a girl, not a boy who wants to be a girl. Unfortunately for me, I was cursed with some physical characteristics that don't match my identity as a girl" (Gorman 2016).

The Parliamentary Friendship Group for LGBTIQ Australians, co-chaired by Warren Entsch, M.P., and Senator Janet Rice drafted a proposal that aims to enable transgender youth access to stage two hormone treatment without applying to the Australian Family Court. Both Entsch and Cathy McGowan, M.P., acknowledged Isabelle (Entsch and McGowan 2015), with McGowan claiming on her website

> Isabelle is a 12-year-old girl who was born a boy. She is a brave and courageous young person who was able to tell her family that she is a girl born in a boy's body. And her family's response was simply and lovingly: "How can we best support Isabelle? How can we create a future that she can live in and thrive in?"
>
> (McGowan 2015)

In her constituency statement to the House of Representatives, McGowan claimed:

> Last week the Australian Human Rights Commission released its report *Resilient individuals: Sexual orientation, gender identity & intersex rights*. It calls for options other than a Family Court order for access to hormone treatment. This is urgent for young Isabelle and transgender children just like her.
>
> (McGowan 2015, 2)

As in the case of Georgie Stone, the instance of a high-profile political figure acknowledging the social rights of a transgender student presents a compelling portrait of the school and community simultaneously functioning as an ideological landscape in which the child sustains and proactively generates individual identity as a form of cultural heritage. The stories of Briella Carmichael, Isabelle Langley, and Georgie Stone, as well as initiatives such as Wear It Purple, Minus18, the Australian Safe Schools Program, and YGender, among others, clearly support Mathy's (2003, 327) claim that

> Considered together, androgyny, bisexuality, and *transgender* identity, as well as intersexuality, have increased our awareness that gender, sexual

orientation, and sex exist along continua that cannot be cleanly or clearly dichotomized into binary opposites without diminishing the genuine diversity of humanity.

Ecotone: transition spaces of possibility

In the discipline of ecology, the transitional zone or junction zone between two or more diverse communities represents an *ecotone* (Dash and Dash 2009, 221). Defined by American ecologist Frederic E. Clements in 1897, the term characterizes the borderline where the courses of interchange or of opposition between adjoining communities can be seen (Clements 1905, 334; Lévêque 2003, 179).

Here, we are suggesting that movements such as "Wear It Purple Day," "Briella Day," and other events that incorporate transgender heritage into the school ecology may be understood as curricula occurring within the ecotone between two communities. The ecological term for this area is the "edge effect." On the one hand, not only do ecotones commonly reveal biological wealth superior to that of either of the communities they divide (Steiner 2002, 28) but the "edge effect" also represents a rich and dynamic site of unique diversity. According to Agarwal (2008, 227):

> An ecotone is a transition between two or more communities. ... It may be considered as a junction zone or tension belt. In extent the *ecotone* is usually narrow, that is, it occupies a smaller area as compared to the areas occupied by the adjoining communities. The communities of the ecotone area commonly contain many of the organisms which are characteristic and often restricted to the ecotone area itself. The frequency and density of some of the species is often greater in ecotone than in the adjoining communities. There is therefore, a tendency for increased variety and density at the community junction, and this is generally known as the *edge effect*.

In social terms, Figure 8.4 aims to represent the ecotone as a site of tension that generates greater diversity. Here, the ecotone includes features of each community that coalesces those features into a new site of cultural heritage. By this model, I aim to apply Guattari's (2008) proposal to "think 'transversally' " (29).

In Figure 8.4, the overlapping of Community A and Community B represents the transition zone, or "ecotone." Consider Community A as the school ecology, for instance, and Community B as transgender students within that landscape. Within the transition zone there exists a dynamic co-mingling, an exchange (intellectual, cultural, emotional, ideological) between the sexuate bodies characteristic of each community. Events such as "Wear It Purple Day," and "Briella Day," among others, signify narrative practices that both literally posit ecological

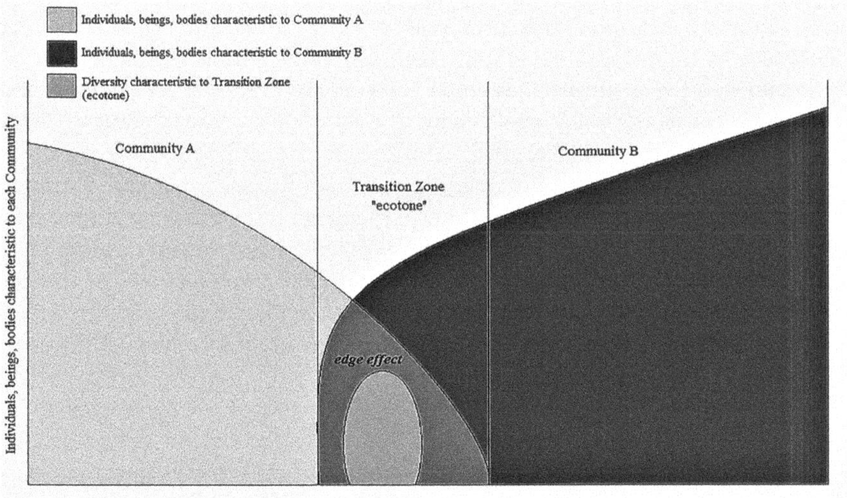

Figure 8.4 Ecotone and edge effect (adapted from Clapham 1973).

meaning through the transgendered child's body—"sexuate difference" (Irigaray 2008, 77)—while simultaneously functioning as a metaphor that defines the subjectivity of Otherness in an educational enterprise of cultural and *human heritage*. Thinking "transversally" can be seen in the rallying around transgender issues of school ecological consequence. There exists here a greater potential for diversification than is possible within the discrete communities on either side. Thinking "transversally" characterizes an affinity, a responsiveness through which the diversity within the ecotone erupts in a coalescence known as the "edge effect." Within both communities there are clearly individuals, beings, and bodies characteristic of those communities, and the possibilities for affinity produced by the ecotone simultaneously create what is known in ecological terms as "edge" species. For instance, while it is highly difficult for transgendered students to exist on the edges of society, their chances for greater well-being rely on the ecotone possibilities formed via the counterpart community. Changes in gender affiliation, for instance, lead to fragmentation within communities while also creating more ecotones. In this respect, the landmarking of transgender with events of cultural heritage can signify, metaphorically speaking, the ecological fact that ecotones are highly dynamic, changeable, and evolving, and lead to greater richness and diversity within human communities. This concept of ecotones harmonizes with Stacy Alaimo's notion of "trans-corporeality" as a "contact zone" (2008, 238) between individuals and the environment which are continuously enmeshed. The edge effect in this concept of ecotones highlights the implications of this inseparability of the human and the "more-than-human nature" (Dowling and Power 2016, 297) in the landmarking of transgender cultural heritage.

Conclusion: moving forward

This chapter has argued that preservation helps assess the significance of transgender cultural sites as heritage and that landmarking places of historic transgender interest further promotes ecological practices that shape and inform the social, cultural, intellectual, and emotional interpretation of these ecological sites as site of "wonder." Transgender sites of ecological preservation represent spaces where discovery is still possible, and where the narrative practices that posit ecological meaning through bodies—"sexuate difference" (Irigaray 2008, 77) and the subjectivity of Otherness—remain ambiguous. They are spaces of wonder that afford memory and remembrances to and for characters not fully developed or even firmly definable; a space within which the transgender individual becomes the person they always knew themselves to be (Poe 2011, 122).

In the discourse of cultural heritage, then, the transsexual body problematizes the concept of "objects of human origin" precisely because, in the politics of identity, these sexuate bodies stress the "intimacy between perceptual subjects (people) and perceptual objects (heritage)" (Kearney 2009, 211). The expression of gender and sexual ambiguity by a single actor/actress moving from one biological "sexing" to another—within and between ecological settings of contemporary interest—opens up rich possibilities for considering the concept of human/cultural heritage. The cultural products of heritage are sites of coalescence and exchange, and in this, schools and school ecologies clearly have a crucial role to play in stimulating transgender awareness and promoting inclusion. Events such as "Wear It Purple Day," "Briella Day," the Safe Schools Program, and websites such as Minus18 and YGender, among others, offer examples of the possibilities for convergence as much as the flourishing richness of ecotones within and beyond the school ecology. These "edge effects" illustrate that just as landmarking is an enterprise of ecological preservation, so too is landmarking "a form of consent, a compact about what is collectively valued" (Sorkin 2011, 132). In fact, these "edge effects," as transsexual heritage enterprises of memory and public recognition, perhaps most significantly, link bodies to social ecologies in practices of transsexual heritage that both inspire and embody human and cultural heritage more broadly. This chapter has theorized about the intersection between transgender youth and young people and place through the lens of cultural significance:

> *Cultural significance* is embodied in the *place* itself, its *fabric*, *setting*, *use*, *associations*, *meanings*, records, *related places* and *related objects*. Places may have a range of values for different individuals or groups.
>
> (Australia ICOMOS 2013, Article 1.2, "Definitions," emphasis added)

By adopting a definition of the term "cultural significance as synonymous with cultural heritage significance and cultural heritage value" (Australia ICOMOS 2013, 2), we have investigated the links between the transsexual body as "related object" and an ecology that "contributes to the cultural significance of a place but is not at the place." Our aim has been to consider institutional—specifically, school—responses to transgender issues and the "discursive practice" (Hall 2005, 25) of heritage that define and affect school ecologies in order to argue that the narrative practices that advance ecological meaning through bodies—"sexuate difference" (Irigaray 2008, 77)—continue to be crucial in the globalization of cultural and *human heritage*.

Notes

1 **Intentional**: For example, in a "Safe School," educators may aim to create a safer and more inclusive environment for same-sex-attracted, intersex, and gender-diverse students.
Curricular: For example, in a "Safe School," educators may adopt a curriculum conducive to creating a safer and more inclusive environment for same-sex-attracted, intersex, and gender-diverse students.
Pedagogical: For example, in a "Safe School," educators may inject their own personality, passions, strengths, and intentions into a curriculum conducive to creating a safer and more inclusive environment for same-sex-attracted, intersex, and gender-diverse students.
Structural: For example, in a "Safe School," educators may dedicate time in the classroom, recruit external partners, and/or observe special days (e.g., "Wear It Purple", etc.), in delivering a curriculum conducive to creating a safer and more inclusive environment for same-sex-attracted, intersex, and gender-diverse students.
Evaluative: For example, in a "Safe School," educators may use observation or informal (and formal) methods of assessing the effectiveness of a curriculum conducive to creating a safer and more inclusive environment for same-sex-attracted, intersex, and gender-diverse students.

2 **Natural and built landmarks** refer to the physical elements (e.g., of the school) perceived through the five senses (taste [gustation], touch [somatosensation], sight [vision], smell [olfaction], and hearing [audition]).
Historical landmarks refer to both the literal and figurative signs of the past (e.g., school buildings, gardens, campuses, school districts, etc.).
Collective landmarks refer to the built objects encouraging community value (e.g., school plaques, memorials to alumni, dedication sites within and beyond school grounds, etc.).
Human landmarks refer to the interpersonal and corporal exchanges with others (staff, students, alumni, and community, both within and beyond the school) that inspire landmarking.

References

ABC. 2016a. "About a Girl." Transcript. *Australian Story* [Television documentary]. *ABC*, August 15. www.abc.net.au/austory/content/2016/s4515734.htm.

ABC. 2016b. Families with Transgender Children Gather in Canberra to Lobby for Hormone Treatment Law Changes. *ABC Radio Canberra*, February 22. www.abc.net.au/news/2016-02-22/parents-of-trangender-children-meet-with-politicians-in-canberra/7189454.

Agarwal, Shyam Kishor. 2008. *Fundamentals of Ecology*. New Delhi: A.P.H. Publishing Corporation.

Abreu, Roberto, L. and Alicia L. Fedewa. 2016. "Home–School Partnerships in LGBTQ-parent Families." In *The SAGE Encyclopedia of LGBTQ Studies*, edited by Abbie E. Goldberg, 538–41. Thousand Oaks, CA: Sage.
Alaimo, Stacy. 2008. "Trans-corporeal Feminisms and the Ethical Space of Nature." In *Material Feminisms*, edited by Stacy Alaimo and San J. Hekman, 237–64. Bloomington: Indiana University Press.
Australia ICOMOS 2013. *The Burra Charter: The Australia ICOMOS Charter for Places of Cultural Significance*. Burwood: Deakin University. http://australia.icomos.org/wp-content/uploads/The-Burra-Charter-2013-Adopted-31.10.2013.pdf.
Australian Government Department of Education and Training 2018. "Safe Schools Coalition Australia (SSCA)," *The Foundation for Young Australians*. Accessed August 18. www.safeschoolscoalition.org.au/who-we-are.
Bailey, Megan. 2016. "Transgender Cranbourne Child Briella Carmichael, 6, Was Born with a Male Body, But Lives as a Girl." *Leader*, February 14. www.heraldsun.com.au/leader/south-east/transgender-cranbourne-child-briella-carmichael-6-was-born-with-a-male-body-but-lives-as-a-girl/news-story/4249e9d00c9b27c5c20f058b6b8a0013.
Behrendt, Poul Olaf and Hansen Per Krogh. 2011. "The Fifth Mode of Representation: Ambiguous Voices in Unreliable Third Person Narration." In *Strange Voices in Narrative Fiction*, edited by Per Krogh Hansen, Stefan Iversen, Henrik Skov Nielsen, and Rolf Reitan, 219–53. Berlin: Walter de Gruyter GmbH.
Bendery, Jennifer and Christian Farias. 2017. "Trump Didn't Actually Roll Back Any Legal Protections for Transgender Kids." *Huffington Post*, February 24. www.huffingtonpost.com.au/entry/trump-transgender-students-bathrooms_us_58af1591e4b0780bac2725d4.
Boske, Christa. 2015. "Bullied: What's Going On?" In *Students, Teachers, and Leaders Addressing Bullying in Schools*, edited by Christa Boske and Azadeh Osanloo, 1–22. Rotterdam: Sense Publishers.
Brans, Edward H.P. 2001. *Liability for Damage to Public Natural Resource, Standing, Damage and Damage Assessment*. The Hague, London, New York: Kluwer Law International.
Brodal, Tina. 2015. "Community Furore after Eight-year-old Who was Born a Boy but Whose Parents Say She Identifies as a Girl was Allowed to Compete in School Run as a Female." *Daily Mail*, June 7. www.dailymail.co.uk/news/article-3113566/Community-furore-eight-year-old-born-boy-parents-say-identifies-girl-allowed-compete-school-run-female.html.
Brown, Leo. 2016. "Safe Schools: Chest Binding Photo Removed from Christian Website after Complaints by Young Transgender Man Shown." *ABC News*, February 26. www.abc.net.au/news/2016-02-26/safe-schools-chest-binding-photo-removed-from-christian-website/7202772.
Chappell, Sharon Verner, Karyl E. Ketchum, and Lisa Richardson. 2018. *Gender Diversity and LGBTQ Inclusion in K-12 Schools: A Guide to Supporting Students, Changing Lives*. New York: Routledge.
Clapham, Wentworth B. 1973. *Natural Ecosystems*. New York: Macmillan.
Clements, Frederic E. 1905. *Research Methods in Ecology*. Lincoln, NB: University Publishing Company.
Cobb, Edith. 1977. *The Ecology of Imagination in Childhood*. New York: Colombia University Press.
Cohen, Janine. 2016. "Transgender Teenagers 'Risking Lives' Buying Hormones on Black Market." *ABC News*, August 15. www.abc.net.au/news/2016-08-15/transgender-teens-buying-black-market-hormones/7722084.

Cohen, Janine and Catherine Scott. 2014. "Being Me." *Four Corners*, November 18. www.abc.net.au/4corners/stories/2014/11/17/4127631.htm.

Dash, Madhab Chandra and Satya Prakash Dash. 2009. *Fundamentals of Ecology* (3rd edn). New Delhi: McGraw Hill.

Davison, Kate. 2011. "Agents of Social Change? LGBT Voices in Australian Museums." In *Queen City of the South: Gay and Lesbian Melbourne*, edited by Graham Willett and John Arnold. *The LaTrobe Journal*, 87(May): 149–63.

Department for Education and Child Development. 2016. *Transgender and Intersex Support Procedure*. Government of South Australia, November. www.decd.sa.gov.au/sites/g/files/net691/f/transgender-and-intersex-support-procedure.pdf.

Devine, Miranda. 2017. "An Epidemic of Transgender Children is Safe Schools' Legacy." *Daily Telegraph*, April 16. www.dailytelegraph.com.au/rendezview/an-epidemic-of-transgender-children-is-safe-schools-legacy/news-story/085d5681f6bc3dae2357302ab-2bee227.

Dierschow, Rebecca. 2016. "Faro and the LGBT Heritage Community." In *Who Needs Experts?: Counter-mapping Cultural Heritage*, edited by John Schofield, 93–100. London and New York: Routledge.

Donnelly, Kevin. 2016. "Criticising Safe Schools Doesn't Make You Homophobic." *ABC News*, March 24. www.abc.net.au/news/2016-03-24/donnelly-criticising-safe-schools-doesn't-make-you-homophobic/7272932.

Dowling, Robyn and Emma R. Power. 2016. "Domesticities." In *The Wiley-Blackwell Companion to Cultural Geography*, edited by Nuala C. Johnson, Richard H. Schein, and Jamie Winders, 290–304. West Essex: Wiley Blackwell.

Drouin, Jennifer. 2014. *Shakespeare in Quebec: Nations, Gender and Adaptation*. Toronto: University of Toronto Press.

Eisner, Elliot W. 1998. *The Enlightened Eye: Qualitative Inquiry and the Enhancement of Education Practice*. Upper Saddle River, NJ: Prentice Hall.

English Heritage. 2008. *Conservation Principles: Policies and Guidance for the Sustainable Management of the Historic Environment*. Swindon: English Heritage.

Entsch, Warren. 2015. "Time to Reaffirm Support for Transgender Community." *Media Release*, November 26. www.warrenentsch.com.au/Media/MediaReleases/tabid/73/ID/907/Time-to-reaffirm-support-for-Transgender-community.aspx.

Foundation for Young Australians. 2016. "Safe Schools Coalition." www.safeschoolscoalition.org.au/who-we-are.

Geidner, Chris. 2016. "Obama Says Transgender Students Are Vulnerable and Must Be Protected." *Buzzfeed*, May 17. www.buzzfeed.com/chrisgeidner/obama-says-transgender-students-are-vulnerable-and-must-be-p?utm_term=.lh5Yk9VXJ#.ptoP5WB0K.

Gorman, Ginger. 2016. "Fighting for Isabelle: The Battle for Transgender Kids to Get Access to Cross-sex Hormones." *News.Com*, February 23. www.news.com.au/lifestyle/parenting/kids/fighting-for-isabelle-the-battle-for transgender-kids-to-get-access-to-crosssex-hormones/news-story/1083395b247646e0db8b6888a5d8334e.

Guattari, Félix. 2008. *The Three Ecologies*. Translated by Ian Pindar and Paul Sutton. London: Continuum.

Hall, Stuart, ed. 2005. *Modernity: An Introduction to Modern Societies*. Malden, MA: Blackwell.

Hill, Robert J., ed.. 2006. *Challenging Homophobia and Heterosexism. Lesbian, Gay, Bisexual, Transgender, and Queer Issues in Organizational Settings*. Hoboken, NJ: Wiley Periodicals.

Irigaray, Luce, ed.. 2008. *Luce Irigaray: Key Writings*. London: Continuum.

Jones, Rachel. 2015. "Re-reading Diotima: Resources for a Relational Pedagogy." In *Re-Imagining Relationships in Education: Ethics, Politics and Practices*, edited by Morwenna Griffiths, Marit Honerød Hoveid, Sharon Todd, and Christine Winter, 1–23. West Essex: Wiley Blackwell.

Kearney, Amanda. 2009. "Intangible Cultural Heritage. Global Awareness and Local Interest." In *Intangible Heritage*, edited by Laurajane Smith and Natsuko Akagawa, 209–225. Abingdon: Taylor & Francis.

Kohlbacher, Sonia. 2015. "Let Your Transgender Kids Live True, Say Parents." *The Australian*, June 9. www.theaustralian.com.au/news/nation/let-your-transgender-kids-live-true-say-parents/news-story/63179951336dde06dd169d890acb9306.

Laussen, Jim. 2016. "Why I Support Safe Schools as a Principal." *ABC News*, February 25. www.abc.net.au/news/2016-02-25/laussen-why-i-support-safe-schools-as-a-principal/7199280.

Lévêque, Christian. 2003. *Ecology: From Ecosystem to Biosphere*. Plymouth, UK: Science Publishers.

Liu, Fei. 2011. "Intangible Heritage." In *Heritage, Memory and Identity*, edited by Helmut K. Anheier and Yudhishthir Raj Isar, 314–17. London: Sage.

Mac Alpine, Dan. 2016. "Ipswich Students Advance Equality and Gender Identity." *Wicked Local*, May 31. http://ipswich.wickedlocal.com/news/20160531/ipswich-students-advance-equality-and-gender-identity.

McGowan, Cathy. 2015a. "Transgender—Remove Obstacles to Show Support." *Cathy McGowan*, June 17. www.cathymcgowan.com.au/transgender.

McGowan, Cathy. 2015b. "Transgender Children Speech." *Commonwealth of Australian, Parliamentary Debates. House of Representatives Federation Chamber Constituency Statements*, June 17. http://parlinfo.aph.gov.au/parlInfo/genpdf/chamber/hansardr/d6700ba7-bfbc-40cf-a2aa-1e339ff1e843/0231/hansard_frag.pdf;fileType=application%2Fpdf.

McLeish, Michael. 2014. "Healthier and Safer Ways to Bind your Chest." *Minus18*, August 18. https://minus18.org.au/index.php/resources/sexuality-info/item/441-how-to-bind-your-chest.

Mathy, Robin M. 2003. "Gay and Lesbian Studies." In *Encyclopedia of Human Ecology: A–H*, Vol. 1, edited by Julia R. Miller, Richard M. Lerner, Lawrence B. Schiamberg, and Pamela M. Anderson, 323–8. Santa Barbara, CA: ABC–CLIO.

Middleton, Karen. 2016. "What Happened to the Safe Schools Program." *The Saturday Paper*, March 26. www.thesaturdaypaper.com.au/news/politics/2016/03/26/what-happened-the-safe-schools-program/14589108003049.

Minus18. 2016. "Australia's Largest Network for LGBTI Youth." Accessed August 18 2018. https://minus18.org.au/.

Murphy, Ann V. 2007. "Beyond Performativity and Against 'Identification': Gender and Technology in Irigaray." In *Returning to Irigaray: Feminist Philosophy, Politics, and the Question of Unity*, edited by Marie C. Cimitile and Elaine P. Miller, 77–92. Albany: State University of New York Press.

Nabham, Gary Paul and Stephen Trimble. 1994. *The Geography of Childhood: Why Children Need Wild Places*. Boston, MA: Beacon Press.

New South Wales Teachers Federation. 2012. "Wear It Purple Day." *News and Media*, August 29. www.nswtf.org.au/news/2016/04/26/wear-it-purple-day-7-september.html.

New South Wales Teachers Federation. 2015. "Rainbow Day–Moruya High School." *GLBTI Newsletter*, February. www.nswtf.org.au/pages/rainbow-day-moruya-high-school.html.

Nicholas, Lucy. 2016. "Calling Sexual and Gender Diversity 'Contentious' Further Marginalises Students." *Swinburne University of Technology*. "Safe Schools Review Findings: Experts Respond. Analysis for The Conversation by Lucy Nicholas, Lecturer in sociology, Swinburne University of Technology," March 21. www.swinburne.edu.au/news/latest-news/2016/03/safe-schools-review-findings-experts-respond.php.

Nicholson, Larissa. 2014. "When the Dress Fits: Carrum Schoolgirl Leaves Gender Definitions Behind." *The Age*, June 6. www.theage.com.au/victoria/when-the-dress-fits-carrum-schoolgirl-leaves-gender-definitions-behind-20140606-39osp.html.

Peters, Jeremy W., Jo Becker, and Julie Hirshfield Davis. 2017. "Trump Rescinds Rules on Bathrooms for Transgender Students." *New York Times*, February 22. www.nytimes.com/2017/02/22/us/politics/devos-sessions-transgender-students-rights.html.

Poe, Danielle. 2011. "Can Luce Irigaray's Notion of Sexual Difference Be Applied to Transsexual and Transgender Narrative?" In *Thinking with Irigaray*, edited by Mary C. Rawlinson, Sabrina L. Hom, and Serene J. Khader, 111–30. Albany: State University of New York Press.

Ristuccia, Joel M. 2013. "Creating Safe and Supportive Schools for Students Impacted by Traumatic Experience." In *Supporting and Educating Traumatized Students: A Guide for School-based Professionals*, edited by Eric Rossen and Robert Hull, 251–64. New York: Oxford University Press.

Roberts, Louise. 2015. "Children Should Be Allowed to Decide Their Gender." *Daily Telegraph*, June 11. www.dailytelegraph.com.au/rendezview/children-should-be-allowed-to-decide-their-gender/news-story/1e25af4103ed20740f8be2a7bc171d2e.

Rummler, Klaus. 2014. "Foundations of Socio-cultural Ecology; Consequences for Media Education and Mobile Learning in Schools." *Medien Pädagogik* 24: 1–17. www.medienpaed.com/globalassets/medienpaed/24/rummler1407.pdf.

Salamon, Gayle. 2010. *Assuming a Body: Transgender and Rhetorics of Materiality*. West Essex and New York: Columbia University Press.

Schlamb, Carmen. 2017. "On the Practice of Narrative Landmarking: Navigating an Ecological Identity Through Self-study." In *At the Intersection of Selves and Subject: Exploring the Curricular*, edited by Ellyn Lyle, 41–52. Rotterdam, Boston, MA: Sense Publishers.

Schofield, John. 2016. "Heritage Expertise and the Everyday: Citizens and Authority In the Twenty-first Century." In *Who Needs Experts? Counter-mapping Cultural Heritage*, edited by John Schofield, 1–12. London: Routledge.

Schwab, Gail. 2016. "Creative Inter-sexuate Inter-subjectivity in the Classroom?" In *Engaging the World: Thinking after Irigaray*, edited by Mary C. Rawlinson, 135–53. Albany: State University of New York Press.

Sergiovanni, Tom. 2000. *The Lifeworld of Leadership: Creating Culture, Community, and Personal Meaning in Our Schools*. San Francisco, CA: Jossey Bass.

Sergiovanni, Tom. 2005. *Strengthening the Hearthead Leading and Learning Together in Schools*. San Francisco, CA: Jossey Bass.

Smith, Elizabeth, Tiffany Jones, Roz Ward, Jennifer Dixon, Anne Mitchell, and Lynne Hillier. 2014. *From Blues to Rainbows: The Mental Health and Well-being of Gender Diverse and Transgender Young People in Australia*. Melbourne: The Australian Research Centre in Sex, Health, and Society. www.beyondblue.org.au/docs/default-source/research-project-files/bw0268-from-blues-to-rainbows-report-final-report.pdf?sfvrsn=2.

Sorkin, Michael. 1991. *Exquisite Corpse. Writing on Buildings*. New York: Verso.

Sorkin, Michael. 2011. *All Over the Map: Writing on Buildings and Cities*. New York: Verso.

Speer, Susan A. 2010. "Pursuing Views and Testing Commitments: Hypothetical Questions in the Psychiatric Assessment of Transsexual Patients." In *Why Do You ask? The Function of Questions in Institutional Discourse*, edited by Alice F. Freed and Susan Ehrlich, 133–58. New York: Oxford University Press.

Steiner, Frederick R. 2002. *Human Ecology: Following Nature's Lead*. Washington, DC: Island Press.

Strickland, Steven. 2014. "To Treat or Not to Treat: Legal Responses to Transgender Young People." Association of Family and Conciliation Courts Fifty-first Annual Conference, *Navigating the Waters of Shared Parenting: Guidance from the Harbour*. Toronto, Canada, May 28–31.

Sunday Night. 2016. "Safe Schools Are Saving Lives." *Sunday Night*, October 16, https://au.tv.yahoo.com/plus7/sunday-night/a/32914039/safe-schools-are-saving-lives/.

Tando, Darlene. 2016. *The Conscious Parent's Guide to Gender Identity: A Mindful Approach to Embracing Your Child's Authentic Self*. Avon, MA: Adams Media.

Tomazin, Farrah. 2016. "Why Safe Schools Program Must Be Kept." *Sydney Morning Herald*, March 12. www.smh.com.au/comment/why-safe-schools-program-must-be-kept-20160310-gng9qc.html.

TSER. 2016. "Why Transgender People Need More Visibility." *Trans Student Educational Resources*. www.transstudent.org/transvisibility.

UNESCO. 2003. *Convention for the Safeguarding of the Intangible Cultural Heritage*. Paris. Accessed August 18, 2018. http://unesdoc.unesco.org/images//0013/001325/132540e.pdf.

Wade, Matthew. 2016a. "Channel 7 Slammed by Trans Advocates for its Safe Schools Coverage." *Star Observer*, October 17. www.starobserver.com.au/news/national-news/channel-7-safe-schools/153381.

Wade, Matthew. 2016b. "The Struggles Trans Kids Face in Australia." *Star Observer*, June 10. www.starobserver.com.au/news/national-news/the-struggles-for-trans-kids-in-australia/149993.

Waters, Stacey, Donna Cross, and Therese Shaw. 2010. "Does the Nature of Schools Matter? An Exploration of Selected School Ecology Factors on Adolescent Perceptions of School Connectedness." *British Journal of Educational Psychology* 80: 381–402. www.ncbi.nlm.nih.gov/pubmed/20109276.

Wilson, Caroline. 2015. "Towards a Thinking and Practice of Sexual Difference: Putting the Practice of Relationship at the Centre." In *Re-imagining Relationships in Education: Ethics, Politics and Practices*, edited by Morwenna Griffiths, Marit Honerød Hoveid, Sharon Todd, and Christine Winter, 23–37. West Essex: Wiley Blackwell.

Wilson, Tim. 2015. "Gays and Parents Both Have Rights." *Human Rights Commission*, Australia, August 27. www.humanrights.gov.au/news/opinions/gays-and-parents-both-have-rights.

YGender. 2017. "Ygender is a Peer Led Social Support and Advocacy Group for Trans/Gender Diverse Young People." Accessed August 18, 2018. www.ygender.org.au/.

Young, Emma. 2015. "Perth MP Claims Transgender Children 'Will Grow out of It'." *WAToday*, June 10. www.watoday.com.au/wa-news/perth-mp-claims-transgender-children-will-grow-out-of-it-20150608-ghj75f.html.

Zamora, Michael. 2016. "Transgender Athlete Reflects on His High School Cross-Country Experience." *The Des Moines Register*, April 8. www.desmoinesregister.com/videos/sports/high-school/2016/08/04/88276362/.

Zandvliet, David, ed. 2013. *The Ecology of School*. Rotterdam, Boston, MA, and Taipei: Sense.

Zara, Christopher. 2015. "Gay Pride 2015: Stonewall Inn, Historic LGBT Site, Awaits Landmark Protection in Pro-development NYC." *International Business Times*, June 21. www.ibtimes.com/gay-pride-2015-stonewall-inn-historic-lgbt-site-awaits-landmark-protection-pro-1976189.

9 Transgender

An expanded view of the ecological self

Gail Grossman Freyne

Ecofeminist philosophy requires that we constantly theorize two recursive practices. On one hand, we need to understand that the way in which we relate to nature will determine how we relate to each other. On the other, the way in which we relate to each other will shape the ways in which we interact with our environment. This endlessly repeating pattern presupposes that we clarify first what we mean by nature and then, equally importantly, refine what we mean by human identity.

What is nature?

A discussion of the concept of nature most frequently takes place within the discipline of science, the area of study that we understand to be more engaged than any other with the exploration of the natural world. We think of nature as out there, separate from ourselves, our background and environment, and our resources. We think of science as the objective method that we use to understand nature. We think of both nature and science as gender free, yet there is nothing gender free about either category when we try to define it.

From a feminist standpoint we can identify what has been named the "science-gender system." Here, the ideologies of gender and science mutually inform each other and then function in our social arrangements to produce an allegedly objective, dispassionate, and male science, which has traditionally made no room for any subjective, emotionally engaged exploration of the world around us. Within this schema, decisions about what is worth studying, which pieces of data will be considered significant and which irrelevant or less useful, will all depend upon the worldview of the scientist making the judgment. We have failed to acknowledge that the so-called objective, scientific experiment that relies on logical proof, verification, and replication never takes place until a series of prior subjective evaluations and decisions have been made.

Likewise, in psychotherapy the therapist's bias will determine what information should be included and what should be left out or overlooked, a practice that will crucially define the ensuing conversation and therapeutic outcome. The scientist will talk of objectivity and the therapist of neutrality where there can be neither. We can never leave our values outside the consulting room or the laboratory. As Evelyn Fox Keller reminds us in *Reflections on Gender and Science*, "It is through these day to day practices that the selection of preferred descriptions and the dismissal of less congenial ones take place; this is where the truly subversive force of ideology makes itself felt" (Fox Keller 1985, 11).[1] The point is that when we delve deeper into the concept of nature it is a good idea to do so accompanied by a healthy dose of self-criticism.

Across the centuries we have asked the question: What is nature? All our answers have depended upon how we deploy ourselves in relation to the rest of the planet. At this point in human history the answer at first seems obvious. I look out of my window in the southwest of Ireland and, filtered by the sea mist, see fields and fuchsia, sea and sky, beach and boulder. This is the commonplace answer: that nature is the sum total of physical reality separate from humankind. However, a little reflection demonstrates that this answer is inadequate. We think of beehives as part of nature but not our own homes, even when they are made of wood and slate; we think of homeopathic medicine as natural, even though it is produced in the same factory as other chemical compounds. The boundary between what is natural and what is not is constantly blurred. Even as these daily, commonsense distinctions must be made they do not take into account the fact that humans are made of the same substances as the rest of the natural world. This fact makes any radical separation between society and nature unintelligible and therefore ultimately untenable.

The alternative perspective, that nature is the sum total of reality, including humans, has ancient origins from Genesis to Aristotle. When we conceive of nature in such cosmological terms as the totality of being, humanity is neither opposed to nature nor separable from it. This definition provides the starting point for much eco-philosophy. This is especially true of deep ecology, the subset that regards human life as just one equal component of a global ecosystem. But once again, while this definition is as philosophically useful as the first is pragmatic, a little reflection uncovers a serious weakness. Total immersion in nature makes it conceptually impossible to deal with human difference. We need to be able to explain our rationality and self-consciousness, those human attributes that set us apart from other animals.[2] Indeed, we have sometimes been tempted to believe that the further we remove ourselves from nature the more rational we become.[3] Nonetheless, proponents of natural law theory assume that only nature can act as our source of morality and fount of wisdom.

From these two contrary perspectives, namely that of total exclusion or total inclusion, we can only conclude that the concept of nature is notoriously unstable. This instability has arisen because every attempt at a definition of nature is to serve some purpose of the one who is doing the defining. Every time the idea of nature is called into service to bolster our own positions—for example, to maintain that homosexuality or the development of stem cells is unnatural—we can see how little objectivity attaches to the concept.

In the light of this ambivalence, when we approach the question of human identity, perhaps conceptions of the self are better addressed within a third definition that understands nature to be that which is essential to a being. This perspective forms a bridge. Starting with this definition, we can acknowledge that the rational and moral aspects of our human selves are of the essence of our species being. While this marks our distinction from the rest of nature it does not preclude us from seeing ourselves as part of a continuum with nature in our physical composition. This is equally true for the transgender and the cisgender person.

Perhaps the phrase "human nature" itself sums up the dilemma: we are both human and part of nature. Yet, if the phrase sums up the problem, does it not also contain within it a neat articulation of the solution?

What is human nature?

There are two questions that endlessly appear in philosophy. In fact, they might be the two questions that underpin the entire project of human reflection, including ecofeminist philosophy. The first question is: What is human nature? Who am I? The second question, which we will deal with later, is: What is the best form of social organization, what is the best way for all of us to live together? Our answer to the first question will shape our answer to the second.

Because these two questions involve a study of both nature and science, it is not surprising to find that the answers that have been given over the centuries have been anything but objective and gender free. Let us examine the problem of human nature through the lens of the third definition of nature that I have suggested which proposes that nature be understood as the essence of our being. One could be forgiven for assuming that if we are dealing with the species "human" the essence would be the same in every individual. For example, I have already suggested that the rational and moral aspect of the human person is of our essence and it is these attributes that distinguish us from other vertebrates.

Unfortunately, at this point in the history of Western thought, despite the best efforts of feminist philosophers, this third definition remains bogged down by the unexamined assumptions inherent in the notion of a dual anthropology, a theory of two human types.[4] This is far from being a new idea but it exhibits a cockroach-like persistence. Across the ages, from the Platonic dualism of reason/nature, to its most recent articulation by the late pope, now St. John Paul II, we are provided with statements on the complementarity and collaboration of men and women,[5] which assure us that "he" is like this and "she" is like that. "He" is dispassionate reason and aligned with the workings of the public sphere while "she" is the reproductive world of the lactating body and the chaotic emotions, and is confined to the private sphere. He is *essentially* more rational and she is *essentially* more caring. It is easy to see which one is superior. This might be considered an overstatement but it is a position that is still responsible for the fact that each sex is underrepresented in both the public and the private spheres of human activity respectively.

These two types of human are not simply different, because their arrangement is also hierarchical. It starts with blue and pink blankets and never falters until we are certain that rational, independent, and assertive men are entitled to higher pay for the same work and that the provision of maternity, rather than paternity or parental leave, is responsive to the fact that only she, as the emotional, dependent, and passive mother is responsible for the children. By overlaying our sex differences with the social construction of gender we are left with two incomplete versions of human nature: the masculine and the feminine. It is not feminism but this binary system of gender identification that has given birth to the war between the sexes. Each sex continues to be defined in opposition to the other. The more each sex behaves in accordance with the gender-role stereotype they have been handed at birth, the further they are pushed apart, one from the other.

Sex differences are used as the justification for gender role stereotyping, which is another way of saying that our human essence, far from being one, has been bifurcated. This is an excellent example of biased self-interest using "nature" to justify the desired form of social organization. Caught within the limiting confines of such a structure no human being, no man or woman, can ever become *fully* human. This binary and rigid description is clearly unsatisfactory for the majority cisgender community, but it pushes the LGBT minorities toward conceptual invisibility. It is a refusal to recognize that while our human essence is one, it has many manifestations. Furthermore, while it is a position that is philosophically flawed, it is also scientifically incorrect. A binary and rigid division into two sexes and two genders does not correspond to what is known about our natural context, where bodies and behaviors in multiple species are in a constant state of fluidity.

If a rigid and binary view of sex differences has been used as a justification for demanding stereotypical gender role behavior, then a more fluid and flexible appreciation of human sexuality will undermine this demand. We could say, for example, that men could be free to become good mothers, sometimes, and women could be good breadwinners, sometimes. Or all the time or none of the time, or whatever they chose. That is to say that every single person would be not only free but expected to manifest the full range of human behaviors.

The other major problem with a bifurcated understanding of human nature is that it fails to correspond to what we have learned from science, not only in terms of species behavior but also with regard to our chemical inheritance. We now know, beyond any doubt, that our species, one among the millions of extant species, evolved out of the same stuff. There is nothing poetic or metaphorical about the statement that our essence is stardust; it is scientific truth. However, instead of focusing on what we share with each other and the rest of nature, we limit our focus to our intra human difference within a fog of philosophical bias, scientific ignorance, and willful anthropocentrism. This is a world in which men are superior to nature, and because women are aligned with nature then they are inferior to men. At first glance, we can see that the trans man immediately becomes clothed in the power and privilege that has traditionally adhered to all men. Yet the trans woman pays a price: she will lose this status of power and privilege the moment she identifies as a member of the traditionally 'inferior' second sex. Trans men and trans women do not come equally to their new gender roles.

What we do not have is a definition of human identity that not only corresponds to all observable reality and, far from being bifurcated, also ensures a *common essence for all* human beings, allowing each person to manifest the full range of human behaviors, virtues, and ethical responses.

The ecological self

Ecofeminist philosophy has addressed itself to the problem of human identity by simply asserting that any definition of our identity must this time correspond to all observable reality. Despite what science tells us about the impact of humans on the rapid deterioration of our planet, we, particularly the male of the species, have never had any trouble asserting how rational we are. Once, we had a more realistic awareness of our place in the cosmos but since the so-called Enlightenment the hubris of modern, anthropocentric man finds it difficult to accept that we are just one of a variety of animals that live on this planet.[6] Collectively, we ignore the fact that we are embodied beings who do not simply walk upon the earth but are deeply embedded in our ecosphere. What are the implications of this deep embeddedness?

As I mentioned at the outset, the way in which we relate to nature is a foundational premise because it will determine the way in which we relate to each other. If men are equally connected to nature, then it can no longer be a ground for the subjection of women. While we recognize our continuity with nature, we also recognize our difference. The trick is in the balance. Attempting to achieve this balance, the ecofeminist philosopher addresses the notion of interdependence by investigating the ways in which human persons and nature depend on each other. How are we akin to nature and how do we differ?

Traditional theoretical attempts to define human identity have not provided the answer. To attempt to posit a new model of what it means to be human, one that accords with all observable and scientific reality, one that is more complex and inclusive than the traditional "Master Model of Humanity" (Plumwood 1993, 5),[7] is the primary interest of ecofeminist philosophy. In other words, there can be no equality between human beings in a model of humanity that has been formed by men to the exclusion of women and nature.

Deep ecology does not provide the answer. We have already seen that deep ecology's refusal to acknowledge any difference in reality between the human and the natural world, the position diametrically opposed to the traditional one, throws up the insurmountable, conceptual difficulty of making it impossible to theorize human difference. In addition, this conceptual maneuver of totally merging the human self with nature throws up a corresponding problem: it fails to recognize the distinctness and independence of every earth other. This is a familiar form of anthropocentric arrogance that fails to recognize the intrinsic value of each living thing. If we fail to recognize the intrinsic value of the other as "other" we have imposed ourselves, our will, our decisions, our view of good and bad, right and wrong, upon them. You can only love, befriend, and care for another if you respect the independent aspect of their being (Freyne 2006, 77).

Feminist philosophy does not provide the answer. Other forms of feminism, in line with traditional, patriarchal thought, have relied on the insistence that women are just as rational as men to achieve equality. The corollary to this path toward salvation entails that they eschew the concept of nature, seeing it as the seat of women's oppression. They have been content to analyze the problems of the "relational self," as he or she interacts with other humans. It is the ecofeminist who also needs to know how we interact with the natural world.[8] This is a deployment of the concept of "self" that is designed to move beyond the limited "relational self" of intra-human interactions to the more encompassing "ecological self." Ecofeminists do accept the definition of nature as the sum total of reality, including humans, but they see this not as the last word but as the best *starting point* in the construction of human identity. It is the recognition of our kinship with nature, as well as our rational difference from nature, that marks ecofeminist thought as different from other feminist philosophies.

The ecological self is a complete synthesis of both parts of the human person, the mind and the body, each part this time holding equal value. Ecofeminist philosophy insists that we must continually analyze the concept of nature so that both parts of the human person, the rational and the emotional, can be brought to bear on the construction of an inclusive, and this time complete model of human identity. To live well with each other, to aspire to the best form of social organization, we must learn to think with our feelings. A good example of this is mourning for the death of a loved one. We engage collectively in the farewell ritual of a funeral and we understand that this leads to social cohesion while the other part of us feels the pain. So it is with the death of the matriarch in an elephant herd, where we see that burial practices are remarkably complex. Her family surrounds her body and lightly touches it with their feet and trunks. Family members cry out and weep. The group eventually covers her with leaves and dirt and stays beside her for days.[9]

The ecological self is a concept that is powerful in both theory and practice. Previous definitions of human identity that have been both anthropocentric and androcentric are upended. First, the ecological self is no longer exclusively human centered. Human beings cannot exist without clear air and fresh water. Like infants, we profoundly depend upon nature's bounty and so as adults must care for the earth. On the other hand, despite our depredations of the earth, ultimately the planet is not dependent upon us for its very existence. Autonomy is an illusion.

When we know and experience our vulnerability, we can start to rethink the previous, hierarchical model of intra-human relationships. From this standpoint the apparently powerful are, in fact, totally dependent. There is no industrial magnate without a workforce and no jobs are created without seed capital. If there are no jobs, there is no money for retail shopping. If the shop owners do not turn a profit, their mortgages do not get paid. Then the homeless need the government to rehouse them, which forces the government to raise taxes on the factory owners. The wealthy tell the unions they cannot afford to give the workers a pay rise, at which point the workers strike and close the factory. We all live, as the ecologists are so fond of saying, within a web of interlocking relationships. Any successful form of social organization will be grounded in the notion of interdependence, not of power. The belief that we have power over any person or system is dangerous for everyone. Just because the rich and powerful will be the last to feel the effects of scarcity of clean air and water does not mean that they will never feel them. Freedom and security are illusions.

Second, we are forced to consider how we are like nature and how we differ. We are all part of evolutionary nature and biology loves complexity. This awareness allows us to reflect upon the implications of our individual differences and the intrinsic value of each one. The human species also embodies multiple variations. We have different skin colors and different religious beliefs, or none at all. We are rich and poor, athletic or physically challenged, some can see, some cannot. Some are gay and some are straight, some are cisgender and some are transgender. And we come in multiple combinations and their expressions are limitless. Yet, we all share a common humanity as rational and ethical beings who can reflect upon and alter our habitat.

In summary, the ecological self recognizes our essential kinship with nature and our human difference from nature and admits that we can never deny our dependence upon nature. As a result, s/he embodies changed attitudes to, and thus behaviors toward, nature. Now that we have resituated humankind in ecological terms, our task is two-fold: how to reconfigure intra-human relationships in ethical terms and how to apply this learning to our relationship with nature.

The natural phenomenon of transgender and cisgender

The second recursive practice to consider here is the one that claims that the way we relate to each other will determine the ways we relate to our environment.

The way we relate to each other will be based on what we have learned from reappraising our relationship with nature: our elemental kinship with and difference from it, as well as our dependence upon it. We have seen that we like our dealings with each other to be "rational" and "natural," a view that corresponds with the notion that "men are like this" and "women are like that." However, our ecological selves are far more complex than we have habitually thought, a fact that Joan Roughgarden, biologist and ecologist, so aptly describes as our evolutionary rainbow.[10] For the biologist and the layperson the formation of a new human being comprises very different understandings. "To a biologist, *'male' means making small gametes and 'female' means making large gametes.* Period!" (Roughgarden 2004, 23, emphasis in original). For most of us, our sex education consists of men producing sperm (small gametes) and women producing eggs (large gametes) and when the two meet we have the start of cellular division that finally results in a new person. But for the biologist, "men" and "women" are social categories, not biological ones.

Social categories, like gender, are open to definition and redefinition, construction and deconstruction. We have the freedom to decide who counts as a man and who counts as a woman and the criteria for this decision, both internal and external, change from time to time. For example, biology tells us that internally some human males do not have a Y chromosome and externally some do not have a penis. Unfortunately, far too recently, medical science and culture prescribed that doctors should surgically provide female genitals to babies who were born without them so that, raised as females, they could conform to an apparently fixed binary system of sexual difference. At the time it was meant as a kindness. Today, there is a growing consensus in the medical world that diverse, intersex bodies are normal—if relatively rare—forms of human biology.[11] They are the children of nature and this diversity is a naturally occurring variation, just as it is in the rest of the animal world. It is not a medical condition.

Furthermore, Roughgarden would argue that the biggest error of biology today is the uncritical assumption that the gamete-size binary (small sperm, large egg) implies a corresponding binary in body type, behavior, and life history (2004, 26). Certainly, she agrees that gender is most commonly understood to refer to a person's biological sex. Most of us have even extended gender to animals, describing aggressive bulls and gentle cows, controlling stallions and compliant mares. But we recognize this only as the behavior of two sexes. For someone like me, arriving late to the science of biology, Roughgarden's contention that many species have three or more genders, with individuals of each sex occurring in two or more forms, is startling. She grounds this contention by suggesting a wider definition of gender. For her, "*Gender is the appearance, behavior, and life history of a sexed body*" (Roughgarden 2004, 27, emphasis in original).

About appearance or body type, she reminds us that in the animal world some females have a penis (female spotted hyenas have a penis-like structure externally identical to that of males) and some males lactate (the male fruit bat of Borneo and Malaysia). About behavior, species demonstrating male parental care are too numerous to mention. In many species the female deposits the eggs in the pouch of the male, who incubates them until birth. Sex-role reversal in nature is commonplace, with males and not females tending the nest. With respect to life history, in approximately half of the animal kingdom the body form of an individual is both male and female at the same time, or at different times during its life. As a layperson, newly apprised of this extraordinary degree of diversity, I am tempted to borrow the words of W.S. Gilbert to exclaim, "Here's a how-de-do! … Here's a pretty mess!"[12] At the age of 70, the words "sex" and "gender" have taken on whole new meanings for me.

What we can conclude is that a rigid, binary division of two sexes exhibiting corresponding masculine and feminine behavior is definitely not the whole story. The unavoidable corollary is that relations between male and female humans are not limited to a complementary relationship; we are all moral agents equally embodied with reason and embedded in nature. We are all vulnerable at certain times in our lives and we deny our dependence upon each other and our environment at our peril. Simultaneously, we are all independent, unique, and intrinsically valuable.

To ground the discussion let us consider the transgender person. The ecological self allows for the visibility of the transgender person in a way that the former, binary definition of human identity did not. This ecological self is constructed in accordance with all observable reality: humans are kin yet different in nature but utterly dependent upon nature. How different would our communities be if we understood transgender persons to be the fulcrum for a more fruitful discussion of human identity?

To postulate transgender persons as central sounds unrealistic. After all, this type of human person makes up about 1 percent of the species. And yet, although they are the minority, could we not posit that they are the most complex and inclusive form of the human person? They are male and female, man and woman, masculine and feminine, sometimes they cross from one gender to the other more than once, and sometimes they are gay and sometimes straight. They are a special synthesis of mind and body, of reason and nature, living with a neurological condition rather than a psychological one. As so many of them express it, they have a mind trapped in the wrong body. Should their special knowledge allow us to consider them normative? If the rest of nature is fluid, then why would the human species, comprised of cisgender and transgender, intersex and androgynous, insist on rigidity? This is not to suggest that the rest of us need to alter our body shape. Rather, the person who is transgender makes it possible for us to think in terms of a unified anthropology, one in which all aspects, attributes, virtues, and potential of the human being exist, either concurrently or sequentially. Allowing transgender persons to be the fulcrum forces us to believe what we have previously refused to acknowledge. First, that humans, like other species, come in many forms. Second, we must locate ourselves on a continuum with other species. In the light of this, our populations would now be more correctly described as composed of transgender and cisgender persons.

To provide evidence for this contention, Roughgarden discusses the lack of foundation for a simple, binary gender system by providing a most interesting analysis of Darwin's theory of sexual selection.[13] She suggests that Darwin has produced the first theory of gender with his observation that it is almost universal to find that males have stronger passions and that females are more coy. As I outlined at the start of this chapter, this appears as a classic case of the scientist's bias infecting his science: Darwin lived at a time when the behavioral codes for men and women were profoundly divisive. In his theory, Darwin postulates that males compete to eliminate the weak and the sickly among themselves. The female then welcomes the strong male so that her offspring will have the best genes. For Roughgarden this theory takes a regressive stance in that it incorrectly views gene pool diversity as consisting of mainly bad genes that males must eliminate and females avoid. At its core, the theory is diversity-repressing within a species, although not across species.

It appears that Darwin knew that many animals do not align with a simple sex binary because he made numerous observations of the male of the species having up to three genders and the female two. Even when distinct male and female bodies existed, he was also aware of the display of sex-role-reversal behaviors.

Sexual selection theory requires that sperm transfer be for reproduction. However, there is a hundred to a thousand times more mating, both heterosexual and homosexual, which proves that mating behavior is as much orientated towards relationships as it is focused on conception alone. The extensive evidence of same-sex sexuality between vertebrates demands that we describe the behavior as anything but an aberration. Yet, it would appear that in this aspect of his work, facts that did not fit Darwin's theory were not given equal standing. As a result, there was no need to explain them other than as "exceptions" (Roughgarden 2004, 171). How many exceptions must there be before the theory is judged suspect or unhelpful?

The importance of this critique is that it shows a multiplicity of bodies and lifestyles existing in the natural world, not only within species but between species. As Roughgarden expresses it (2004, 182), nature is not diversity repressing:

> nature offers a smorgasbord of possibilities for how to live,
> and an endless list of solutions for every context. …
> The true story of nature is profoundly empowering for peoples
> of minority gender expressions and sexualities.

If political scientists and sociologists were to study the relationships on display in the natural world and allow these as models for intra-human relationships, then profound possibilities for change would be free to emerge.[14] Immediately, it would be possible to reconfigure both elements of the phrase "human nature" simultaneously. We could recognize that transgender people are willing, indeed feel compelled, to break the most serious "human" taboo of all: to express the Other within themselves. They do not engage in mimicry, they become not just like the Other because they already are the Other. They show us how to change what we always thought was fixed. We are now provided with a new understanding of our relationship with the natural world. Instead of limiting it as a mere resource from which we take without counting the cost, a means to human ends, the rest of nature now becomes our *partner* in evolution, demonstrating that every configuration of the ecosphere, every living creature in the biosphere, has independent and intrinsic value in its own right.

Transitioning to the ecological self

Time is running out for humans to make dramatic changes in the ways we relate to our planet to ensure our survival. It is now urgent that we radically reform our notions of a partial and separated human identity and take on all the implications that flow from being ecological selves.

To transition, to move from one state to another or to a different self-understanding, is not something that only a few humans will do—it is imperative for all of us. Dissolving the reason/nature dualism is central to this universal task of transitioning. In the history of Western thought reason has reigned supreme; only reason was the source of all human knowledge. It was coded male and it despised the lower and chaotic world of nature and emotion that was coded female. Nature was the world of women, children, barbarians, and slaves, and it was far too corrupt to be the source of real knowledge. But it is not the difference between every living thing that has been the problem. The problem arises only when difference is co-opted and arranged in hierarchical, dualistic form. The solution is to understand that "different from" does not mean "better than" or "less than," it simply means different. Thus, the way to dissolve a dualism is to affirm both sides of it simultaneously. All humans think and feel, but we cannot be fully human until we ascribe equal value to both activities. As we learn to think with our feelings we can become, finally, fully human. Indeed, the only fully human person, be they transgender or cisgender, is the one who has transitioned to the ecological self.

The first step of the transition is to admit that we have been wrong about a lot of things. Not just about Platonic dualisms but also about tying a child's left hand behind her back so that she will be forced to write with her right hand. We have been wrong about homosexuality. Likewise, we have been wrong about transgender persons, again believing that therapy would "fix" them. These errors spring from a compulsion to fear and denigrate difference.

As I have argued, a dual anthropology is the result of refusing to give primacy to what we all share, our *common good*: our rational, emotional, and moral selves. Instead, the focus was upon difference so that women's bodies, and gay bodies, and transsexual bodies, were aligned with nature and therefore represented a difference that was "less than." Today, the study of biology tells us that we would be wrong to keep insisting that the gamete-size binary found in humans corresponds to a binary in human body type, behavior, and life history. When we admit that we have been in error across centuries, from the denial of Galileo to the acceptance of gayness, we have arrived at the essential element of self-criticism that allows us to deploy ourselves ethically in relationship to our environment from a position of openness. As the ecofeminist philosopher Val Plumwood expresses it: "Our capacity to gain insight from understanding our social context, to learn from self-critical perspectives on the past and to allow for our own limitations of vision, is still one of our best hopes for creative change and survival" (2002,10).

In particular, in our social context, we admit to the blind prejudice that has precluded us from being able to see value in the Other. For example, the vast majority of us never contemplate the egregious suffering of animals bred and farmed for profit. But to treat another with respect, sympathy, sensitivity, and consideration we need to engage in a process of ethical reflection. According to Iris Young, we must step back from our own impulses and desires in order to consider them in relation to the just demands of others (1991, 105). We know that we must limit self-interest, that the needs of one may have to await the flourishing of another. But having our own standpoint—and, as I have already argued, we cannot be without one—must work hand in hand with the virtue of empathy which, in turn, requires that we sometimes adopt the standpoint of the other.

As we work to make visible the existence of the transgender person, we begin to validate their existence. We eliminate the stigmatizing diagnosis "gender identity disorder" and replace it with the phrase "gender dysphoria." Gender reassignment surgery, hormone treatment, and counseling are made available so that each person can be who they are, or who they understand themselves to be, rather than forcing them to conform to the sex assigned to them at birth. Anti-discrimination laws are passed, as are laws to allow the transgender person into the workforce and its bathrooms. Educators are being taught to deal with the early emergence of gender dysphoria in young children. We are changing our relationship to the transgender person just as we did with the homosexual person.

Simultaneously, a process identified by Nicole Seymour (2017), which she terms organic transgenderism, may be taking place. This is a process that focuses on transitioning not within the Western medical model but "as a self-directed, even spontaneous phenomenon akin to the life-cycle changes of plants and animals."[15] Most interestingly, she sees this process as something that emerges from an expanded ecological consciousness or, as I would see it, as transitioning to the ecological self.

We have also been in error in believing that what is natural and normative are descriptions limited to straight men and straight women. This leads us into the more complex and therefore potentially more fruitful world of LGBT relationships. A movement toward complexity makes it very clear that we all exist on a continuum and a much more interesting one than we have heretofore imagined.

It is the transgender person who is the epitome of human biological diversity. In their identity, they embody a complexity (noted above) that we find in other animals as they transition from one state to another, a parallel that clearly places all of us on the same human continuum but also on a continuum with nature. We have done our best to ignore this continuity with our environment, adopting a philosophical and ethical stance of closure rather than one of openness. This has led relentlessly to the denial of our dependence upon nature, to a refusal to acknowledge the intrinsic value of every earth other while at the same time asserting human superiority. There is no need to list examples here of where such thinking has led us. Oceans of ink have been spilled on the topic. But the transgendered among us are pushing out the boundaries of human identity. In so doing they illuminate the path to dissolving the human/nature dualism and give us one last chance to renegotiate our relationship with our environment.

Second, this pushing the boundaries of human identity forces the recognition that "human" is not a complete or even sufficient description of our "nature." We are also primates and vertebrates. This is an infinitely more comprehensive self-description and, as we have seen from the work of Roughgarden and countless biologists, it immediately places us far beyond the binary gender system. All human beings are part of the genus human. If we insist on using the less inclusive descriptions like heterosexual, straight, or cisgender as being the only acceptable moral categories, then we leave other categories outside of ethical reflection except to deny them ethical value. If we fail to accept transgender people as fully and perfectly formed human beings, then the cis community limits the possibilities of what they themselves could become. We repeat the original error that was the construction of the master model of humanity. If we leave any human out of consideration, then our final definition of human identity will again be distorted as it was when we attempted to construct it without consideration of women and nature. It precludes the formation of properly integrated selves.

Third, a fully integrated self cannot emerge from the theoretical perspective of a dual anthropology. Half of all human potential and virtue is denied to the other half. Why should a human male, of his essence, not be gentle, passive, intuitive, and the primary caregiver of his children? Why should a female, of her essence, not be decisive, active, and coolly rational as the primary breadwinner for her family? Masculine and feminine are adjectives. We can use these adjectives to describe various forms of behavior but they do not describe any fundamental or definitive reality. More than that, far from gender role stereotyping, we can attribute them at different times to different bodies.

Our first question was "Who am I?" It is a question that contains another question, "Who was I designed to be?" These questions are asked by everyone, sometimes only once, sometimes repeatedly over the course of a life. The questions are asked because they contain cultural undertones of an ethical value: What is acceptable? What do I need to do to conform? Am I a real man? Am I a good woman? The transgender person is asking these questions on a level far more profound than most of us will ever reach. But they can teach us how to reach the parts of ourselves that we continually deny in an effort to conform to someone's else's idea of what we should be. It's the "should" that gives the game away, exposing the ethical, socially constructed, requirement.

World of nature, your name is variety. Among humans, it is the transgender person who embodies the greatest degree of variety and from this place of complexity they show the rest of us how to be who we think we should be and who we feel we would like to be. Many of my clients have struggled with questions of identity. One woman was a lawyer who wanted to be married but never wanted children. Her husband, family, and friends pressured her to become a mother. Another was a gay man who married because he was too fearful to come out even to himself. Then there are the famous, like Group Captain Catherine "Cate" McGregor, a transwoman who was once Malcolm McGregor. She tried hard to conform but eventually realized that the macho aggression she exhibited was masking something much deeper. She felt that "A death wish was driving me." It takes courage to face losing your marriage ("the love of my life"), your family, your friends, your job, your home, especially when you have just been granted the Order of Australia for being someone else. But the Group

Captain has transitioned and serves in the Australian Defence Force with 15 other transgender people. She and her wife have remained best friends, the head of the Army refused her resignation, and she is still a cricket writer.[16]

The point is that we all continually struggle with our identity even when we do not realize it. We take nature for granted, it is simply the air that we breathe and we never think about it. Unless there is a problem, unless there is pollution. We do the same thing with our sexual identity, never reflecting upon it because we have never suffered mis-gendering.

It is transgender people, like Cate McGregor, who rattle our self-satisfaction. And I know of one little boy in Australia who likes to dress as a girl sometimes. When one of his playmates irritably asked, "Are you a boy or a girl?" he replied, perhaps not yet knowing, "Both!" This child shows that autonomy, freedom, and security are perhaps still possible; s/he is the embodiment and carrier of these dreams. As Naomi Scheman has said, marginalized lives "are lived, and hence livable" (1996, 132).[17] These lived lives cannot and should not be denied. But personal conversion or individual enlightenment is not enough. Nature, transgender, and cisgender people need institutional protection in the form of laws. Even if we experienced personal conversion at the private level and passed laws for equality and non-discrimination at the public level, not enough would change. You only have to ask the suffragettes or the second-wave feminist movement for confirmation. Talk is cheap; reformers do a lot of it. In the end, structural change will depend on cultural disobedience, which is why we need courageous people like Cate McGregor, who will "chain themselves to the rails" to attract the attention of the powerful, the policy-makers. Transgender people still have the power to shock, because their very existence is a form of cultural disobedience. And therein lies their power. Why locate the body as the problem? Surely the problem still lies, as it has always done, in our attitudes to and practices around the body.

We have come to accept left-handed people and homosexuals, and in time the transgender person will become as boringly commonplace as their cisgender sisters and brothers. But until that moment arrives they are the ones who prove the biological diversity of the human person, our own species rainbow, which only mirrors the variety to be found in the natural world. To socially organize around that reality, rather than one of closed and centric hubris, will enable us to move toward a new standpoint in relationship to our environment as the human mind is educated by nature.

Notes

1 For an expansive discussion of the relationship between gender and science and how objectivity itself comes under suspicion as an androcentric goal, see Evelyn Fox Keller (1985).

2 Two examples of this type of unnuanced thinking are provided by Robyn Eckersley (1992, 49), and Warwick Fox (1984, 199). For Eckersley, there are no discrete entities and no absolute dividing lines between the animate and inanimate or the human and nonhuman. For Fox, there is no bifurcation in reality between the human and the nonhuman. It is difficult to know what reality is being invoked in either position. It

seems fair to suggest that useful distinctions can be made between particles of sand and human beings.

3 For a persuasive argument detailing the dangers of this worldview see Val Plumwood (2002). She has argued that an illusory sense of independence from nature is irrational and has resulted in multiple forms of ecological denial that threaten both person and planet.

4 Within transfeminist politics this dual anthropology has been defined by Julia Serano as *cisgenderism*, a term she defines as indicating the assumption that males ought to be masculine and females ought to be feminine where masculinity and femininity are constituted by the attributes typically associated with males and females respectively (2007, 90).

5 For a full discussion of the problems with this pope's articulation of human identity see Gail Grossman Freyne (2004).

6 Carolyn Merchant (1980). Merchant contrasts the mechanistic account of nature arising with the Enlightenment with previous organic models of nature.

7 This model of what it means to be human operates to the disadvantage of women *and* men *and* nature.
See Val Plumwood (1993).

8 Some forms of feminism have eschewed the concept of nature, seeing women's alignment with it as the source of their oppression. Other feminists have argued that women are closer to nature than men and therefore better than men, an argument that produces only a reverse chauvinism.

9 Andrew Aghapour (2016).

10 Roughgarden (2004). See also Bruce Bagemihl. (1999).

11 Zederic (2002). Furthermore, it is believed that approximately 1 per 1,000 births are intersex and that 1 per 1,300 XY (genetic male) fetuses develop female genitals without internal reproductive organs (p. 8).

12 *The Mikado*, comic opera by Gilbert and Sullivan.

13 This is not to be confused with Darwin's theory of natural selection.

14 Nicole Seymour (2017, 260) argues that a more justice-orientated perspective on transgender experience requires us to shift from a ciscentric and anthropocentric viewpoint (*most humans are cisgender and therefore cisgenderism must be natural across all life)* to an inclusive, ecocentric viewpoint (*many animals are transgender, transexual, or intersex: how might that change how we think of ourselves as human animals?).*

15 Seymour (2017, 257).

16 Rick Feneley (2014).

17 Naomi Scheman (1996).

References

Aghapour, Andrew. 2016. "What if Animals Believe in God?" *Religion Dispatches*, March 7. California: USC Press.

Bagemihl, Bruce. 1999. *Biological Exuberance: Animal Homosexuality and Natural Diversity.* New York: St. Martin's Press.

Eckersley, Robyn. 1992. *Environmental and Political Theory.* London: UCL.

Feneley, Rick. 2014. "Cate McGregor on Leaving Malcolm McGregor Behind and Becoming a Woman." *The Sydney Morning Herald*, February 25.

Fox Keller, Evelyn. 1985. *Reflections on Gender and Science.* New Haven, CT: Yale University Press.

Fox, Warwick. 1984. "Deep Ecology: A New Philosophy of Our Time?" *Ecologist* 14.

Freyne, Gail Grossman. 2006. *Care, Justice and Gender: A New Harmony for Family Values.* Dublin: Veritas.

Merchant, Carolyn. 1980. *The Death of Nature: Women, Ecology and the Scientific Revolution.* San Francisco, CA: Harper Collins.
Plumwood, Val. 1993. *Feminism and the Mastery of Nature.* London: Routledge.
Plumwood, Val. 2002. *Environmental Culture: The Ecological Crisis of Reason.* London: Routledge.
Roughgarden, Joan. 2004. *Evolution's Rainbow: Diversity, Gender and Sexuality in Nature and People.* Berkeley: University of California Press.
Scheman, Naomi. 1996. "Queering the Center by Centering the Queer: Reflections on Transsexuals and Secular Jews." In *Feminists Rethink the Self,* edited by Diana Tietjens Meyers, 124–62. Boulder, CO: Westview Press.
Serano, Julia. 2007. *Whipping Girl: A Transsexual Woman on Sexism and the Scapegoating of Femininity.* Emeryville, CA: Seal Press.
Seymour, Nicole. 2013. *Strange Natures: Futurity, Empathy, and the Queer Ecological Imagination.* Urbana, IL: University of Illinois Press.
Seymour, Nicole. 2017. "Transgender Environments." In *Routledge Handbook of Gender and Environment*, edited by Sherilyn MacGregor, 253–69. New York: Routledge.
Young, Iris. 1991. *Justice and the Politics of Difference.* Princeton, NJ: Princeton University Press.
Zederic, Stephen. 2002. *Pediatric Gender Assignment: A Critical Reappraisal.* New York: Kluwer Academic/Plenum Publishing.

10 "Good animals"

The past, present, and futures of trans ecology

Nicole Seymour

Who cut me from
  growing into a buck?

Who left me
    only horns and hips?

    Still, I am a good animal.

  Strong,
but not too strong.

Oliver Baez Bendorf, "II"

I teach Oliver Baez Bendorf's poetry collection *The Spectral Wilderness* in two different courses at my university: "Queer Literature and Theory" and "Literature and the Environment." This fact speaks, of course, to my idiosyncratic specialization in queer and environmental studies, and perhaps also to Bendorf's unique perspective as a trans man raised on a farm in the Midwestern United States. But it also speaks to the larger insight that this volume represents: that trans issues are environmental/ecological issues, and that environmental/ecological issues are trans issues.

I have been grappling with this insight for over a decade now. The initial inspiration was a Caribbean literature course in grad school in which my professor, Vera Kutzinski, assigned Trinidadian-Canadian author Shani Mootoo's novel *Cereus Blooms at Night* (1996). The novel weaves a complicated tale in which a traumatized cisgender character named Mala cares for both human and nonhuman lives—including those of a cereus plant and the transgender narrator Tyler—thus stimulating the "blooming" of both. The novel set off a light bulb in my head about the possible connections between LGBTQ issues and environmental/ecological issues; between queer theory and ecocriticism. Searching for scholarship that theorized this connection, I came upon the queer ecology scholarship of Greta Gaard—whose preface graces this volume—and Catriona Sandilands (2005)—whose work is central to I-min Peter Huang's essay (Chapter 5, this volume).[1] The eventual result was my first book, *Strange Natures: Futurity, Empathy, and the Queer Ecological Imagination* (2013), which included a chapter on *Cereus*, Michelle Cliff's novel *No Telephone to Heaven* (1987), and Leslie Feinberg's novel *Stone Butch Blues* (1993). In that chapter, I claim that these works develop a counter-discourse that I term "organic transgenderism" (Seymour 2013, 36): a vision of gender transitioning as spontaneous, immanent, self-driven, and generally "natural"—akin to the life-cycle changes of plants and animals—in opposition to its dominant framing in popular and medical discourse as an "unnatural," technoscientific intervention.

My experience of writing that chapter sparked the recognition that, while queer ecology scholarship has paid attention to non-normative sexuality, it has largely ignored non-normative gender.[2] Thus, through a subsequent series of essays and book chapters over the past few years, I have attempted to model and develop what I have been calling "trans ecology."[3] As a scholar with a background in literature and media, this framework has allowed me to see how trans and environmental/ecological issues intersect in texts ranging from reality TV (Seymour 2015) to road movies (Seymour 2016) to young adult novels (Seymour 2017) to poetry (Seymour forthcoming). And as an interdisciplinary scholar, I have also been particularly interested in developing meta-critical analyses—considering the intellectual genealogies that trans ecology might draw upon, as well as how affiliated areas, including ecocriticism, the New Materialism (including material ecocriticism and feminism), queer ecology, transgender studies, and environmental justice studies might more fully address the intersection of trans and environmental/ecological issues.

I will build on that work in what follows, returning intermittently to Bendorf's poem "II" as an illustrative text. First, I will outline what I see as shared impulses across eco and trans frameworks, including an opposition to binaries and a concern with risk and endangerment. As I show, the contributors to this volume build upon those shared impulses and bring them closer together. I will then turn to a survey of issues that I believe trans ecology scholarship should take up in the future, including (re)considerations of aesthetics and of the capacious definition of "trans" found across the humanities. I will conclude with a call for the development of an ethics of trans ecology.

Eco/trans: tracing the convergence

Both eco and trans frameworks have long struggled against binaries, dichotomies, and dualisms. To begin with the first: ecocritics have complicated the distinction between Culture and Nature, at least as far back as Cheryll Glotfelty's field-defining statement in the *Ecocriticism Reader*: "all ecological criticism shares the fundamental premise that human culture is connected to the physical world, affecting it and affected by it. Ecocriticism takes as its subject the interconnections between nature and culture" (1996, xix). Meanwhile, ecofeminists have opposed a much larger list of binaries, as with Val Plumwood's famous schema that includes male/female, mind/body, rationality/animality, civilized/primitive, subject/object, and self/other (1993, 43); Gaard's founding gesture of queer ecology was to add reason/the erotic and heterosexual/queer to this list (1997, 116).

Opposition to the Culture/Nature binary has been crucial to transgender studies as well, as when Susan Stryker reminds us in her Foreword to this volume that "embodiment is always technologized embodiment, that soma and techné never really have an 'and' between them" (p. xvii). While at first such an observation seems to run counter to a concept like organic transgenderism, both share the impulse to depathologize trans embodiment—a point to which I will return later. Further, transgender studies has also critiqued the (cisgender) Male/ (cisgender) Female binary. Indeed, the recent tendency both within and outside the field toward shortening "transgender" to "trans"—discussed in greater depth below—works in part to dispel the notion that all non-cisgender people follow a trajectory from their assigned gender to its "opposite." More pointedly, "trans" can accommodate individuals who explicitly identify as nonbinary.

I believe we can identify this skepticism toward binaries in Bendorf's poem, when we consider its movement from male ("buck") to female ("hips and horns") to something broader and more balanced: animal ("strong, but not too strong"). We might point out here that although bucks are often defined by their antlers, and females of the same species by their absence, those are not universally defining characteristics; for example, female reindeer also have antlers. Importantly, then, the poem makes these seemingly relevant markers of gender give way to the base category of living being. Making "animal" the final defining descriptor of the speaker, I would argue, not only enacts a form of organic transgenderism—naturalizing trans humanity—but also demonstrates a reluctance to reassert the trans individual's rights on the speciesist basis of "humanity."[4] Bendorf's trans-species perspective places, as Straube (Chapter 3, this volume) puts it, "more-than-human entities on equal terms with the human trans body" (p. xvii).

In addition to this shared skepticism toward binaries, eco and trans paradigms have long been concerned with risk and endangerment. For example, ecocritics and other environmental humanists have focused on environmental degradation, endangered animals, and the risky environs that environmental injustice creates for humans. Meanwhile, trans scholars and activists have highlighted the risks to trans individuals, as with projects such as Transgender Day of Remembrance, a global honoring of those who have lost their lives to anti-trans violence.[5] As Stryker writes (Foreword, this volume), "trans-life … is a perpetually precarious life, a life always at risk of death and subjugation" (p. xvii).

At the same time, we must note that some environmentalists have taken threats to the environment as an opportunity to stigmatize trans embodiment. As Huang reports (Chapter 5, this volume), drawing upon the work of Giovanna Di Chiro, "The dominant anti-toxic discourse regularly and typically appeals to patriarchal anxiety about and fear of 'chemical castration,' 'feminization of nature,' 'pan-species *instability* of maleness' and loss of '*natural* masculinity'" (p. 104, drawing upon Giovanna Di Chiro [2010, 201]). But as Malin Ah-King and Eva Hayward have asked, "Why is sex more central than cancer, autoimmune disease, and even death [in conversations around toxic exposure]?" (2013, n.p.). Similarly, I have questioned the assumption "that people suffer from transgender *ontologies*, rather than from *transphobia*, such that we have to get to the root of transgenderism, but not to the root of transphobia—or, say, corporate pollution" (Seymour 2017, 259).[6] Trans ecology, instead, sees risks to the environment and risks to trans persons as parallel or even interconnected. For example, in an essay inspired by the Crochet Coral Reef art project, Jeanne Vaccaro argues that "[v]iolence threatens transgender bodies and coral colonies alike, in registers of diverse feeling and administration as, for example, street harassment, un- and underemployment, toxic waters and chemical pollution" (2015, 286). Anna Bedford's Introduction to this volume captures these insights, summing up: "trans and ecological Others are intertwined in their subjugation within the cisheteropatriarchy" (p. 2).

Bendorf's poem articulates this sense of risk and endangerment, beginning as it does with the specter of bodily violence: "Who cut me…?" (2015, Kindle location 509). I think it is particularly important that the speaker shifts the responsibility for this violence away from himself and onto an external actor; doing so counters popular narratives around gender transitioning as "self-mutilation" and stresses the idea that the real threats to trans people are external. Further, I would point out that Bendorf doesn't adopt just any nonhuman animal persona for this poem, but one that invokes forms of animal exploitation, including hunting and meat-eating ("buck" is a generic term for the male of antlered species, including deer and antelope). Again, we see the idea of the trans human and the nonhuman as equally threatened.

In addition to these specific points of shared concern, eco and trans studies have recently taken broad, complementary turns toward the others' historical purview, thus further paving the way for a trans/eco framework and, of course, this volume. As suggested above, ecocritics and other environmental humanists have become increasingly attuned to queer and trans issues in recent years.[7] Meanwhile, we have seen a growing interest within transgender studies in topics that are, if not always explicitly "environmental," generally relevant to ecocriticism and the environmental humanities. For example, as Jenny Sundén has reported, drawing upon Susan Stryker and Aren Aizura's (2013) Introduction to *The Transgender Studies Reader 2*, "work that blurs the line between human and nonhuman [is] increasingly common in transgender studies, pointing to the presence of trans-animal studies" (2015, n.p.). Indeed, just a month after Sundén published this observation, *TSQ: Transgender Studies Quarterly* released its special issue on "Tranimalities," edited by Eva Hayward and Jami Weinstein; while often highly theoretical, the essays in this issue also addressed "real-world" concerns around environmental topics such as species extinction. (See, in particular, Cleo Woelfle Erskine and July Cole's "Transfiguring the Anthropocene: Stochastic Reimaginings of Human-beaver Worlds" (2015).) Perhaps we can see Bendorf's choice of an animal speaker in "II" as the creative iteration of these intellectual interests.

Transgender studies scholars have also recently shown great interest in questions of space and place—the traditional purview of ecocriticism, environmental history, and other environmental humanities frameworks. Lucas Crawford's recent work on architecture, affect, and transgender experience is one important example. In a brilliant chapter on Virginia Woolf's novel *Orlando*, Crawford calls the titular character's home a "generator of … feeling" (2015, 95), pointing out that it changes throughout the ages just as Orlando's gender does. Indeed, I would argue that Crawford's work is implicitly though not explicitly "environmental," concerned as it is with our emotional and material relationships to our environs, including but not limited to our own bodies. Elsewhere, Crawford has shown how trans coming-out narratives privilege the urban over the rural–a privileging that Katie Hogan (Chapter 7, this volume) critiques. Similarly, rural trans writers and activists such as Eli Clare have described their alienation from urban, supposedly progressive queer spaces (1999). Likewise, Bendorf's *Spectral Wilderness* collection articulates his gender transitioning in relationship to his family farm in Iowa; space and place are inseparable from his understanding of himself as a changing body.

Eco/trans futures

The above genealogy, I hope, provides an account of how something like "trans ecology" or "transecology" could, and has, come about. In this section, I offer a vision of how trans ecology scholars might move forward. I begin with the previously stated point that Bendorf's poem "II" engages in organic transgenderism. Not only does the speaker imagine himself as a nonhuman animal, but the narrative implied by the first two lines—that the speaker *would have* grown into a "buck" on his own, were it not for outside interference—naturalizes his (trans) maleness. We see this kind of naturalization throughout Bendorf's collection, as with the poem "Outing, Iowa," in which the speaker tells us,

> If you've ever doubted that a body can transform/completely, take the highway north from town …/… and go left at the arrow for the lake. Can I tell you? The land where I was/born was born an ocean, and that ocean born of ice. …/… I still bleed, still weep:/what we used to be matters.
>
> (Kindle location 204)

Scholars in this volume have found similar rhetorical processes at work in other texts. For example, Elizabeth Parker argues that David Ebershoff's novel *The Danish Girl* associates its protagonist, a fictionalized version of landscape painter Lili Elbe, with the environment, thus presenting this trans woman as largely natural (Chapter 1, p. 24, this volume). Going forward, then, trans ecology work might consider the limitations and possible exclusivity of what now appears to be a classic trans rhetorical-political move. For example, what of authors, characters, or individuals without access to the rural landscapes explored in Bendorf's poetry or Elbe's paintings? What are the alternative strategies available to those in less idyllic landscapes for naturalizing—or at least de-pathologizing—trans embodiment? How do factors such as race and class shape one's ability to make this rhetorical-political move? And perhaps a more pointed question: To what extent does the naturalization of trans embodiment described here rely on traditional, even conservative ideals around pristine or pastoral nature–ideals that play into the Culture/Nature binary discussed above?

Trans ecology scholars might also undertake historicizing or comparative work, cataloging and/or weighing up the *different ways* in which trans characters, artists, and individuals have appealed to concepts of nature in order to naturalize themselves. Consider, for example, how the butterfly emerging from its chrysalis has been an important emblem for various trans communities, almost to the point of cliché—as seen with a US community publication from the 1990s, *Chrysalis Quarterly* (later, *Chrysalis: A Journal of Transgressive Gender Identities*); a current line of lingerie named Chrysalis, designed for transgender women; and a UK-based counseling center of the same name.[8] Bendorf uses a different emblem of natural transformativity on the cover of *The Spectral Wilderness*: a tadpole at various stages of transformation into a frog or toad. What difference does the choice of animal or natural entity—butterfly, buck, frog, or landscape—make? What does the employment of these various natural entities say about how we, trans and cis humans alike, apprehend the nonhuman?

In addition to the aforementioned questions, trans ecology scholars might also continue to think deeply about aesthetics, perhaps using Huang's (Chapter 5, this volume) work as a guiding example. We might ask: How does it matter whether an artwork that lends itself to trans ecological thinking is a poem, a film, a novel, a television show, or something else? Do particular genres or forms uniquely lend themselves to trans ecological perspectives? How so? I want to briefly model such thinking by considering T.C. Tolbert's questions from the introduction to the collection *Troubling the Line: Trans and Genderqueer Poetry and Poetics* (2013, 11):

> Can a trans and genderqueer poem pass and does it want to? Aside from possible narrative overlap, would there be [unique] syntactic, stylistic, and/or imagistic themes? Do we want that? Is there such a thing as a trans and genderqueer poetics? A trans and genderqueer genre? A trans and genderqueer form?

Tolbert's first question is somewhat difficult to answer because it depends on multiple factors, including the context in which one encounters the poem and the reader's own sensibility. When I teach "II" without first introducing Bendorf's bio, maybe only one-third of my students—usually those in my Queer Literature and Theory course—intuit that the poem is in some way about being genderqueer and/or trans. (I should note here that *The Spectral Wilderness* features a Preface from Mark Doty (2015) that explicitly identifies Bendorf as a transitioning man at the time of its writing; as a whole, the collection is not trying to "pass.") Even if one misses the specific resonances of this poem, though, they would at least notice that its content focuses on corporeal matter ("hips," "horns," "str[e] ng[th]," etc.) and that the form is unusually complex and pointedly stylized.

Indeed, almost all students talk about the two curves found in the poem, from the first to the second line, and from the third to the final.

But regardless of how they initially understand "II," my students and I ultimately come to the conclusion that Bendorf is developing a trans and genderqueer poetic form here. The poem's shape, as at least one student has mentioned, evokes the notion of feminine curves, as in the "hips" it references. We thus talk about how the poem morphologically captures the vicissitudes of a changing body, "lending poetic form to 'a body that has been historically illegible'" (Edwards 2014, 252, quoting Ely Shipley 2013, 197). These vicissitudes, importantly, are not limited to humans but also found in transforming animals like tadpoles, and in sex-changing animals like clownfish. In this sense, we can say that Bendorf is developing a specifically *trans ecological* poetic form. But, as I point out, we could also read the poem's curves affectively and rhetorically. For one thing, the second, larger curve coincides with a move from despair to contentment; beyond a lament over one's feminine body to a self-assured acceptance of oneself as "strong" and "good." The poem's form, in short, is trans-affirming, in all senses of the term.

Poetry seems a particularly ripe area for the kind of trans ecological analysis I have modeled above. As I have shown, poetry can communicate information through form, thus mapping onto trans ontology. And the poetic precedent for adopting other voices can, as I have also shown, articulate transspecies and posthumanist perspectives. Meanwhile, the subfield of ecopoetics has been shifting in promising ways. For example, the premiere ecocriticism journal *ISLE: Interdisciplinary Studies in Literature and Environment* recently published a cluster edited by Angela Hume and Samia Rahimtoola on "queering ecopoetics" (2018); scholars might take advantage of this momentum by continuing to theorize a specifically *trans* ecopoetics. Of course, questions of form and style are not limited to poetry; trans ecology scholars might turn to many other forms and genres to pursue these questions.

But perhaps the largest job for trans ecology scholars would be to grapple with a rather striking convergence between eco and trans frameworks that I have not yet mentioned: how scholars from both areas have developed wildly expansive conceptions of "trans." We could sum up this impulse with the phrase, "Everyone is 'trans' and everything is 'trans' … and yet only some people and some things are trans." Or, to paraphrase Jean Baudrillard's provocative 1987 essay, "We Are All Trans Now?"[9] As I explain in the next section, we find some gaps and complications in how different frameworks pursue this impulse.

We are all trans now?

To begin with transgender studies: in an influential introduction to a 2008 special issue of *Women's Studies Quarterly*, or *WSQ*, Susan Stryker, Paisley Currah, and Lisa Jean Moore proposed "trans-" as a new theoretical concept, arguing that the hyphen at the end "marks the difference between the implied nominalism of 'trans' [no hyphen] and the explicit relationality of 'trans-,' [hyphen] which … resists premature foreclosure by attachment to any single suffix [including gender]" (11). A few years later, in the Introduction to the inaugural issue of *TSQ: Transgender Studies Quarterly*, Stryker and Currah declared that "Although we retain *transgender* in the full, formal title of this journal, we invite you to imagine the *T* in *TSQ* as standing in for whatever version of *trans-* best suits you" (2014, 1). (Interestingly, both of those journal issues featured contributions with eco-adjacent content, from a *WSQ* article by Natalie Corinne Hansen that examines how one man's relationship with his horse is affected by his gender transitioning to a *TSQ* prose piece by Bendorf on the concept of "Nature.") But perhaps the most striking broadening of "trans" has come from Bailey Kier, who, in a *Women and Performance* article on the omnipresence of synthetic hormones in our everyday lives, contends that "everybody on the planet is now encompassed within the category of transgender" (2010, 189).

Meanwhile, material ecocritics and feminists have also been interested in trans phenomena, broadly construed. For example, in her book *Bodily Natures: Science, Environment, and the Material Self*, Stacy Alaimo introduced the concept of "trans-corporeality," telling us that "[i]magining human corporeality as trans-corporeality, in which the human is always intermeshed with the more-than-human world, underlines the extent to which the substance of the human is ultimately inseparable from 'the environment' " (2010, 2). Synthesizing scientific work on, for example, the micro-organisms that live on our skin and in our stomachs, Alaimo shows us that all humans are "trans," insofar as all humans are intermeshed with their environments and therefore not corporeally autonomous. This work chimes with earlier insights from across posthumanism, science and technology studies, and elsewhere, such as the "actor-network theory" made famous by Bruno LaTour starting in the 1980s, which argues that there are no such things as individual objects, only relationships.

These expansive concepts of "trans" have multiple implications. To begin with transgender studies: they stand to do the same work that "queer" has done for queer theory for better or worse. First, they potentially entail a disavowal of identity politics. So, just as queer theorist Lee Edelman has claimed that "queerness can never define an identity; it can only ever disturb one" (2004, 17), transgender studies scholar Aizura argues that "we need a trans theory [that] turns 'trans' in an anti-identitarian direction" (2010, 135). In this sense, "trans," like "queer," functions as oppositional, dissident, and critical, rather than affirmative or biopolitical. Second, an expansive concept of "trans" guards against exclusivity or balkanization, promising this work as relevant far beyond the relatively small demographic of transgender identified persons—just as queer theory has proven itself relevant far beyond the biopolitical formulation of "gays and lesbians."

At the same time, such expansive concepts threaten to render "trans" meaningless in the same way that, say, "queer" has been applied to everything, and thus sometimes means nothing. Perhaps this is why even those scholars who champion a broad gloss of "trans" seem uneasy with that very proposition; after

Stryker and Currah declare that "we invite you to imagine the *T* in *TSQ* as standing in for whatever version of *trans-* best suits you" (2014, 1), they seem to foreclose this freewheeling possibility with a footnote that observes,

> While many of us use "trans studies" in casual conversation, we decided not to call this simply a journal of trans studies, because it either seems too unspecific for general or formal usage or would entail addressing a wide range of trans- phenomena other than those involving gender.
>
> (2014, 17)

So, everyone is trans and everything is trans, except ... only some people and some things are trans. I actually think that might be right—a point I will return to shortly.

Broadly expansive concepts of "trans" also threaten to further elide unique (and uniquely *human*) transgender experiences such as anti-transgender violence and transgender underemployment and homelessness. Further, this expansive turn may, in some iterations, sacrifice a political force or sense of accountability, thus failing to make sense of the human agencies (or the lack thereof) implicated in "real-world" issues. Here, we might think of Richard Kerridge's comment on the nonhuman turn in the humanities, in light of current environmental crises: "there [seems] to be something paradoxical about dispersing and qualifying our notion of human agency at the very moment we need to make an unprecedented demand upon that agency" (2014, 367). Put another way: while it might be important, for various reasons, to recognize that micro-organisms live on our skin and in our stomachs, micro-organisms cannot vote on climate change legislation. Thus, trans ecology scholars should attend to ethical questions such as these: How can we expand concepts of "trans" without emptying that category of its historical, cultural, and political meanings, especially when it comes to the transphobic conditions under which most transgender people live every day? How can we attend to the conditions of social—and, I would add, environmental—injustice under which most transgender people live every day, while not privileging the individual, sovereign human subject? How can we expand concepts of "trans" in ways that explicitly incorporate environmentalist action?

Another set of ethical questions arises when we recognize that, while we might all be trans (that is, porous creatures open to the world) some of us are more trans or open or porous than others, and detrimentally so. To take just one example, according to a recent report from the Public Health Institute and American Public Health Association, climate change threatens to increase heat-related deaths among African Americans, who are more likely to live in inner cities, and who already die from heat at a rate 150 to 200 times greater than non-Hispanic whites (2018, 8). Thus, a theory of trans ecologies would remind us to ask: If everyone and everything is trans, why do only some get punished for it? Why do only some suffer from it? Why are those some (when it comes to humans, at least) overwhelmingly black, brown, poor, and/or female?

Putting "trans" in more direct conversation with categories such as race and class could ameliorate what Zakiyyah Iman Jackson calls the "resounding silence in the posthumanism, object-oriented, and new materialist literatures with respect to race" (2015, 216). And it would remind us not to over-romanticize impurity or pollution. Consider, for example, how Straube's excellent essay (Chapter 3, this volume) celebrates their primary text's elimination of "the dichotomy of clean versus polluted" (p. 54), or how Eva Hayward's important work against purity politics reminds us "that we are already living in ruination" (2014, 258). While these points have potentially liberatory implications for transgender bodies, and while they work against Nature vs. Culture pastoral idealism, it is also the case that pollution has very real, and very negative, effects on the poor and people of color the world over. And it is also the case that academics are largely not living in the same version of ruination as, say, the working poor in the Global South or even the American South. Clearly, then, an ethically and social justice-oriented theory of trans ecologies has immense potential for both scholarly work and activism.

In addition to (human) social justice, this ethical turn would also help us think more carefully through nonhuman relations, or what some scholars have been calling "multispecies justice." As a jumping-off point, let us consider Hayward's short article "Transxenoestrogenesis," which details how both transgender female ontology and cisgender menopausal female ontology are in part enabled by the use of synthetic hormone treatments such as Premarin, which is produced from the urine of pregnant mares—animals that are forcibly inseminated over and over again. Hayward takes this as a paradigmatic example of the transspecies character of human life, briefly acknowledging the ethical upshot: "The cultivation and exploitation of equines has been built into the biopolitics of transwomen" (2014, 256). But what should we *do* with that fact? Can we be both "trans-positive," so to speak, *and* critical, if not condemning, of degradation and exploitation? Returning briefly to Bendorf: Does organic transgenderism, or various forms of self-naturalization such as comparing oneself to animals, always entail positive outcomes, such as, say, animal advocacy? A theory of trans ecologies, I propose, would more deeply explore the range of different transspecies ontologies and imaginaries, and their true implications for the nonhuman.

When it comes to material ecocritics' and other environmental frameworks' expansion of "trans," I have noted elsewhere that "the 'trans' in 'trans-corporeality' does not seem to have any implied resonance with 'trans' as in transgender" (Seymour 2017, 255). In this volume, Straube agrees, pointing out that "Alaimo refrains from embedding the link between trans corporeality and trans embodiment" (Chapter 3, p. 67). Thus, a theory of trans ecologies might put the "trans" as in "trans*gender*" into "trans-corporeality," asking questions like: How is the "trans-ing" experience of a transgender body different from that of a cisgender body? "How might trans identities … be adopted, prompted, or conceived through moves across species or other ontological borders? … And [w]hat are the limits to those processes?" (Seymour 2017, 257), and "In what ways might transgender identities, experiences, or frameworks uniquely lend themselves to recognizing or addressing environmental problems?" (Seymour 2017, 265). Of course, one could go too far in that direction. For example, some scholars have claimed that trans bodies are never bounded bodies. But *no* bodies are bounded bodies, as the scholars cited above—from feminists to scientists—have argued. Indeed, Stryker's aforementioned point about the inseparability of body and technology builds upon Donna Haraway's classic concept of the cyborg, dispelling the notion that only *some* bodies are technologically mediated, while others are natural.

Another problem with materialist approaches, as I have claimed elsewhere, is their very focus on the material body. Transgender studies scholar Gayle Salamon has offered a critique of these approaches, arguing that "a return to the 'real' of the body has troubling consequences … [that] can be seen in the work of several critics of transgenderism"; as Salamon explains, opponents of sex-reassignment surgery "see the phenomenon as evidence that the real of the body is resistant to ideologies of gender" (2006, 587). That is, the stubborn materiality of one's "truly" female or "truly" male body and its biological apparatuses will resist any claims one might make to a different gender identity. Now, certainly, the material body matters to many transgender individuals. I am thinking here of ecologist Cleo Woelfle-Erskine's (n.d.) ruminations on early play in the field, both literally and figuratively: "These were spaces where, alone and without others, except, sometimes, my sister—I had no gender but just a body. And that body was not seen as in a mirror, but rather experienced through touch and contact with dirt grass and water." But we also see in the work of Mootoo, Cliff, Feinberg, Bendorf, and others how the *imagined* body can matter much more, especially in cases where one does not have access to medical technology, or where medical personnel engage in gatekeeping around gender transitioning. Indeed, I would argue that Bendorf's poem represents a process of re*conceiving*, not (just) remaking one's body. Thus, a concept of "trans" that is mainly material may be an incomplete one.

In other words, what we find here yet again are competing impulses. Expanding "trans" threatens to obscure the environmental and other injustices (and other experiences) unique to transgender people, but expanding "trans" also potentially destigmatizes transgender embodiment by showing that it is not a unique experience. "Trans" is material, but "trans" is also immaterial. Everyone and everything is trans, but only some people and some things are trans. My point is not to suggest that some of these impulses are wrong while others are right, nor that we must decide from among them. I simply want to, first, point to a scenario in which we seem to want it all ways, and, second, to propose that one task of a trans ecology framework would be to work through this scenario explicitly, considering how we might choose to balance these impulses, and insisting on a consideration of ethics and justice at every stage possible.

Conclusion

A theory of trans ecologies has yet to be fully developed in the same way that, say, queer ecologies has, though this volume is clearly a major step in that direction. So I want to conclude by summarizing the ongoing functions of trans ecology—or transecology, as this volume's title has it:

- Helping us recognize, and parse, the growing body of artistic work that understands ecological and environmental issues to be inextricable from trans(gender) issues.
- Allowing us to see the intersection among ecological/environmental issues and trans(gender) issues in artistic work that seems to have little to do with either.
- Allowing us to explore the connections between embodied experiences of nature and environment and the idea of coming-into-one's body.
- Continuing to help us assess how transphobia, cisnormativity, and cissexism inform representations and experiences of the more-than-human world.
- Enabling us to recognize, and critique, how discourses around environmental degradation pathologize trans(gender), queer, and otherwise non-normative bodies.
- Enabling us to understand how trans ontologies are, simultaneously, potentially bound up.
  with environmental degradation and animal exploitation and developing a trans eco ethics that would respond to this paradigm.

Thus far, I have conceived of trans ecology as similar to, but distinct from, queer ecology. I have demonstrated how trans ecological scholarship would combine insights from the environmental humanities (including ecocriticism, ecofeminism, queer ecology, environmental ethics and philosophy, and environmental justice studies) with those from an already interdisciplinary field: in this case, transgender studies, rather than queer theory. But considering how the posthuman and nonhuman turns inaugurated in part by environmental humanists have captured the imagination of the humanities at large, considering how regularly the humanities draws insights from the sciences, *and* considering that scholars may be right to say that everyone and everything is trans, then a theory of trans ecologies is no niche inquiry, not a subfield of a subfield, nor (just) the joining of two different areas. A theory of trans ecologies would potentially, and controversially, be a theory of everything.

Notes

1 Queer ecology is sometimes referred to as "queer ecocriticism," though I prefer the former for its expansiveness.
2 As Gaard puts it, "[t]o date, the queer ecocritical focus on sexualities has not captured the critique of heteronormative gender that trans* perspectives address" (Preface, p. xxii, this volume).
3 The earliest use of this term that I have found comes from a January 2014 call for papers for a conference organized by Harlan Weaver and Veronica Sanz. Oliver Bendorf also employs the phrase in his May 2014 keyword article for *Transgender Studies Quarterly*, "Nature." While this collection is obviously titled *Transecology*, I maintain the two-word formulation, both to honor this aforementioned history and to follow the style of "queer ecology."

4 See Hansen for an example of such reassertion; drawing upon a real-life case, Hansen argues that one particular trans man's self-authentication relies on a dominant attitude toward nonhuman animals (2008, 90).
5 For more information, see https://tdor.info/.
6 See also Kier (2010, 314).
7 Arguably, ecofeminists have been concerned with gender since that field's founding in the 1970s—though this has not necessarily entailed queer gender. Even so, as Peter I-min Huang indicates, the potential intersections between ecofeminism and transgender studies deserve our attention (Chapter 5, this volume).
8 See, respectively, www.digitaltransgenderarchive.net/col/2801pg32c, www.chrysalislingerie.com/, and www.chrysalis-gii.org/chrysalis-gender-identity-issues.
9 See "We Are All Transsexuals Now," in *Screened Out*, in which Baudrillard declares, "We are all transsexuals. Just as we are all potential biological mutants, so we are all also potential transsexuals. And this is not even a matter of biology. We are all symbolically transsexuals" (2002, 10). The most interesting part of this essay for me is Baudrillard's reference to La Cicciolina, "the Hungarian-born porn star Ilona Staller, who went on to become an Italian member of parliament on an anti-nuclear, pro-sexual freedom platform" (2002, 10). The connections between anti-nuclear activism and transgender identity are just one area that trans ecology might probe.

References

Ah-King, Malin and Eva Hayward. 2013. "Toxic Sexes: Perverting Pollution and Queering Hormone Disruption." *O-Zone: A Journal of Object-Oriented Studies* 1: 1–12.

Aizura, Aren. 2010. "Transnational Transgender Rights and Immigration Law." In *Transfeminist Perspectives in and beyond Transgender and Gender Studies*, edited by A. Finn Enke, 133–52. Philadelphia, PA: Temple University Press.

Alaimo, Stacy. 2010. *Bodily Natures: Science, Environment, and the Material Self.* Bloomington: University of Indiana Press.

Baudrillard, Jean. 2002. "We are all Transsexuals Now." *Screened Out*. London: Verso.

Bendorf, Oliver Baez. 2014. "Nature." *TSQ: Transgender Studies Quarterly* 1(1–2): 136–7.

Bendorf, Oliver Baez. 2015. *The Spectral Wilderness*. Kent, OH: The Kent State University Press.

Clare, Eli. 2015 (1999). *Exile and Pride: Disability, Queerness, and Liberation*. Durham, NC: Duke University Press.

Cliff, Michelle. 1987. *No Telephone to Heaven*. New York: Vintage International.

Crawford, Lucas. 2015. *Transgender Architectonics: The Shape of Change in Modernist Space*. New York: Routledge.

Di Chiro, Giovanna. 2010. "Polluted Politics? Confronting Toxic Discourse, Sex Panic and Eco-normativity." In *Queer Ecologies: Sex, Nature, Politics, Desire*, edited by Catriona Mortimer-Sandilands and Bruce Erickson, 199–230. Bloomington: Indiana University Press.

Doty, Mark. 2015. "Foreword." In *The Spectral Wilderness*, by Oliver Bendorf, Kindle locations 71–104. Kent, OH: The Kent State University Press.

Edelman, Lee. 2004. *No Future: Queer Theory and the Death Drive*. Durham, NC: Duke University Press.

Edwards, Rebekah. 2014. "Trans-poetics." *TSQ: Transgender Studies Quarterly* 1(1–2): 136–7.

Feinberg, Leslie. 1993. *Stone Butch Blues*. Ithaca, NY: Firebrand.

Gaard, Greta. 1997. "Toward a Queer Ecofeminism." *Hypatia* 12(1): 114–37.

Glotfelty, Cheryll. 1996. "Introduction: Literary Studies in an Age of Environmental Crisis." In *The Ecocriticism Reader: Landmarks in Literary Ecology*, edited by Cheryll Glotfelty and Harold Fromm, xv–xxxvii. Athens: University of Georgia Press.

Hansen, Natalie Corinne. 2008. "Humans, Horses, and Hormones: (Trans)Gendering Cross-species Relationships." *Women's Studies Quarterly* 36(3–4): 87–105.

Hayward, Eva. 2014. "Transxenoestrogenesis." *Transgender Studies Quarterly* 1(1–2): 255–8.

Hume, Angela, and Samia Rahimtoola. 2018. "Introduction: Queering Ecopoetics." *ISLE: Interdisciplinary Studies in Literature and Environment* 25(1): 135–49.

Jackson, Zakiyyah Iman. 2015. "Outer Worlds: The Persistence of Race in Movement 'Beyond the Human.'" *GLQ: A Journal of Lesbian and Gay Studies* 21(2–3): 215–18.

Kerridge, Richard. 2014. "Ecocritical Approaches to Literary Form and Genre: Urgency, Depth, Provisionality, Temporality." In *The Oxford Handbook of Ecocriticism*, edited by Greg Garrard, 361–76. Oxford: Oxford University Press.

Kier, Bailey. 2010. "Interdependent Ecological Transsex: Notes on Re/Production, 'Transgender' Fish, and the Management of Populations, Species, and Resources." *Women and Performance: A Journal of Feminist Theory* 20(3): 299–319.

LaTour, Bruno. 1988. "Mixing Humans and Nonhumans Together: The Sociology of a Door Closer." *Social Problems* 35(3): 298–310.

Mootoo, Shani. 1996. *Cereus Blooms at Night*. New York: Avon Books.

Plumwood, Val. 1993. *Feminism and the Mastery of Nature*. New York: Routledge.

Public Health Institute and American Public Health Association. 2018. "Health Equity and Climate Change." In *Climate Change, Health, and Equity: A Guide for Local Health Departments*.

Salamon, Gayle. 2006. "Boys of the Lex: Transgenderism and Rhetorics of Materiality." *GLQ* 12(4): 575–97.

Sandilands, Catriona. 2005. "Unnatural Passions? Notes toward a Queer Ecology." *Invisible Culture: An Electronic Journal for Visual Culture* 9. www.rochester.edu/in_visible_culture/Issue_9/issue9_sandilands.pdf.

Seymour, Nicole. 2013. *Strange Natures: Futurity, Empathy, and the Queer Ecological Imagination*. Champaign: University of Illinois Press.

Seymour, Nicole. 2015. "Alligator Earrings and the Fishhook in the Face: Tragedy, Transcorporeality, and Animal Drag." *TSQ* 2(2): 261–79.

Seymour, Nicole. 2016. "Trans Ecology and the Transgender Road Trip." In *The Oxford Handbook of Ecocriticism*, edited by Greg Garrard. Oxford: Oxford University Press. E-book edition.

Seymour, Nicole. 2017. "Transgender Environments." In *The Routledge Handbook of Gender and Environment*, edited by Sherilyn MacGregor, 253–69. Abingdon: Routledge.

Seymour, Nicole. Forthcoming. "Queer Ecologies and Queer Environmentalisms." In *The Cambridge Companion to Queer Studies*. Cambridge: Cambridge University Press.

Shipley, Ely. 2013. "The Transformative and Queer Language of Poetry." In *Troubling the Line: Trans and Genderqueer Poetry and Poetics*, edited by T.C. Tolbert and Trace Peterson, 187–98. New York: Nightboat Books.

Stryker, Susan, and Paisley Currah. 2014. "Introduction." *TSQ: Transgender Studies Quarterly* 1(1–2): 1–18.

Stryker, Susan, Paisley Currah, and Lisa Jean Moore. 2008. "Introduction: Trans-, Trans, or Transgender?" *Women's Studies Quarterly* 36(3–4): 11–22.

Sundén, Jenny. 2015. "On Trans-, Glitch, and Gender as Machinery of Failure." *First Monday* 20(4).

Tolbert, T.C. 2013. "Introduction." In *Troubling the Line: Trans and Genderqueer Poetry and Poetics*, edited by T.C. Tolbert and Trace Peterson. New York: Nightboat Books.

Vaccaro, Jeanne. 2015. "Feelings and Fractals: Woolly Ecologies of Transgender Matter." *GLQ: A Journal of Gay and Lesbian Studies* 21(2–3): 273–93.

Woelfle-Erskine, Cleo. n.d. "Queer x Trans x Ecology: Towards a Field Praxis." In *Underflow: Transfiguring Riverine Ecology*. In preparation.

Woelfle-Erskine, Cleo and July Cole. 2015. "Transfiguring the Anthropocene: Stochastic Reimaginings of Human-Beaver Worlds." *TSQ: Transgender Studies Quarterly* 2(2): 297–316.

Afterword

You'd be home—meditations on transecologies

Finn Enke

The rains pour down as I write this, each drop and growing torrent carrying water, dust, ash, jet fuel, radioactivity, hormones. It's been raining for weeks, here where I am. I know how it feels on my skin, and I try to appreciate rain as just rain, these waters that are both life-giving and devastating depending on their location, quantity and timing. But rain isn't what it used to be, and neither am I. Rain is all mixed up, too, as it drops globe-trotting chemicals into this small and now-soggy part of the world while elsewhere forests and towns go up in flames for want of water.

As I write this, nearly 100 skeletons belonging to black bodies have been found in a mass grave unearthed in Texas. Buried post-Abolition, these bodies were people no longer called slaves (by law), but now prisoners forced into the same labor—cotton—on which capitalists had come to depend. Since at least 1905, developers have been unearthing the mass graves and cemeteries of African Americans conscripted to rebuild the state economy through uncompensated, body-breaking labor. Such discoveries are so common at Texas construction sites that the question is almost always where to relocate the bodies, not where to relocate the building project.

At the exact moment my fingers move across the keyboard creating digital figures on a digital page, people are being sentenced to years in prison for trying to protect indigenously held lands and the waterways of the Dakotas from the ravages of a growing number of oil pipelines. Meanwhile, the Lakota Sioux continue to refuse the Black Hills Land Claim—a settlement now topping 1.5 billion dollars to "pay for" the illegal theft of Sioux lands after the 1868 Fort Laramie Treaty had promised the Sioux "undisturbed use and occupation" of a large swath of land that includes the Black Hills.

I'm still writing as perceptible amounts of exogenous testosterone applied to the skin on my forearms slowly becomes endogenous, mixing with sweat, blood, rain, all the molecules of water and not-water that make up this interbeing.

Ecologies, transgenderings, speculative fictions, are all lenses that together invite contemplation of time, place, embodiment, belonging, and home. In the Afterword to this wide-ranging collection, I touch upon additional works with which transecologies might converse, particularly those in areas of indigeneity, dis/ability, and queerness. I return to this morphing, constellated question: What are bodies, and what is home, when we acknowledge the layers of history that we imagine ourselves simultaneously into and out of?

In *Bodyminds Reimagined: (Dis)Ability, Race and Gender Black Women's Speculative Fiction*, Sami Schalk (2018, 17) attends to spaces "in which the rules of reality do not fully apply," spaces that invite us to "imagine otherwise" and reconsider the "meanings and possibilities of bodyminds." People generate and consume speculative fiction and transecologies precisely because the "rules of reality" do not come close to describing our unruly lived realities nor suggest possible alternative survivals. For Schalk, "the rules of reality" refer to "the culturally and historically specific social narratives of the possibilities and meanings of bodyminds, time, space, and technology, as well as our constructed notions of what constitutes a 'real' disability, gender, race, and so on." Who actually lives there? It's a rhetorical question. We imagine otherwise with the very stuff of our present realities; utopian futurities live *now* alongside dystopia and apocalypse (Munoz 2009).

As the chapters in *Transecologies* so clearly illuminate, humans and nonhumans who are marginalized or even exterminated under the conditions of "reality" know that whole communities, time, place, life, and death all exceed conventional definitions of "realness." I'm reminded of Jina Kim's (2017) work elaborating the notion of "dependency" as a critical analytic in the face of racist, capitalist infrastructures. Analyzing Jesmyn Ward's novel, *Salvage the Bones*, Kim reimagines "dependency" as an active constitution of community and agency when the un/natural flooding of New Orleans during Hurricane Katrina magnifies the disparate distribution of resources and dis/ability. We code-switch not only between contexts, but also because multiple realities *co-exist*, each having some "rules of reality" for those who might be considered indigenous to it (Munoz 2012).

My ears stutter a little each time I hear people talk about how the ecosystem in one region is 14,000 years old, in another, 8,000 years old; certain hills are some millions of years old. These datings sometimes suggest the amount of time since the last major cataclysmic event—such as a nearly continent-sized flood or a glacier coming and going—scraped one layer and replaced it with another. How long has it been since an eruption literally turned one part of the planet inside out? We divide things this way.

As a white American traversing and not landing in one purportedly discreet gender, I cannot help but wonder about the intervals. What about all the survivals, all the living as well as all the dying, during those "crisis" times between one "era" and another? How long did it take to build a city on a wetland, and how long will the locks and dams that allow a city to (mis)"manage" a watershed hold? Isn't every period including the present a crisis of transformation, however quickly or slowly it proceeds? As white supremacy, fascism, and capitalism ravage the earth, it's hard not to wonder what life will look like on the "other side" of this slow-/fast-motion cataclysm. What even is a "watershed" anymore?

To say this all takes place on stolen land is too easy; be wary of inclusions that might "control and absorb dissent" rather than admit transformation (Arvin, Tuck, and Morrill 2013, 17).

Engagements with gender and sexuality are engagements with settler colonialism, and they raise questions about what sovereignty and self-determination can mean today and in the future and for whom. As Qwo-Li Driskill (2016, 89) eloquently elaborates in *Asegi Stories: Cherokee Queer and Two Spirit Memory*, "Colonial heteropatriarchy maps gender and sexuality onto Indigenous bodies in order to find routes into and through our homelands." Resistance, then, depends on seeing gender and sexuality as central sites of social transformation. For Driskill, one of the most powerful mechanisms is to weave an "*asegi* imaginary." In the Cherokee language, *asegi* means "strange," and it has been taken up by some as a term of self-identification with the sense of being "strange-hearted," a distinctly Cherokee turn on queerness and the more generalized indigenous concept of two-spirit. *Asegi* imagination, according to Driskill, is vital to the survival and flourishing of indigenous peoples.

Restitution is not exactly return but, like restoration, must paradoxically embrace transformation. In *Brilliant Imperfection: Grappling with Cure*, Eli Clare engages contradictory concepts of dis/ability and "cure" to illuminate ways that bodyminds and the earth have been conditioned by settler colonial definitions of normal/abnormal, natural/unnatural that attempt to parse out who and what belongs where. Clare (2017) writes,

> So you work hard to restore this 8,000 year old ecosystem, all the while remembering that the land isn't yours nor the dairy farmer's down the road, but rather stolen a mere century and a half ago from the Dakota people. The histories of dirt, grass, genocide, bison massacre float here.... The work isn't about recreating a static landscape somehow frozen in time, but rather about encouraging and reshaping dynamic ecological interdependencies, ranging from clods of dirt to towering thunderheads, tiny microbes to herds of bison, into a self-sustaining system of constant flux.

Flux. It's always with us. Perhaps transecologies succeed most of all in making explicit the affair humans maintain in on-again, off-again, hide-and-seek fashion, with flux. It's not all liberatory, either, as Ellen Samuels reminds us in "Six Ways of Looking at Crip Time." While crip time may—much as speculatives and futurisms and otherwises do—encourage us to define our own "normal," it is also broken time, grief time, sick time. Precisely because crip time is one meeting place between mortality and socially constituted rules of normality, crip time is also *time travel*. As Samuels (2017) writes,

> Disability and illness have the power to extract us from linear, progressive time with its normative life stages and cast us into a wormhole of backward and forward acceleration, jerky stops and starts, tedious intervals and abrupt endings. Some of us contend with the impairments of old age while still young; some of us are treated like children no matter how old we get. The medical language of illness tries to reimpose the linear, speaking in terms of the chronic, the progressive, and the terminal, of relapses and

> stages. But we who occupy the bodies of crip time know that we are never linear, and we rage silently—or not so silently—at the calm straightforwardness of those who live in the sheltered space of normative time.

It's not always comfortable, the interbeing of bodyminds and mortality. Yet, as they used to say of Planet Earth, "it's the only home we've got." And it is still vast, beyond our wildest imaginations.

References

Arvin, Maili, Eve Tuck, and Angie Morrill. 2013. "Decolonizing Feminism: Challenging Connections between Settler Colonialism and Heteropatriarchy." *Feminist Formations* 25(1): 8–34.

Clare, Eli. 2017. *Brilliant Imperfection: Grappling with Cure*. Durham, NC: Duke University Press.

Kim, Jina. 2017. "Cripping the Welfare Queen: Disability and Race in the Afterlife of U.S. Welfare Reform." Keynote address presented at Bodies of Resistance Symposium, Department of Gender and Women's Studies, University of Wisconsin, March 2.

Munoz, Jose Esteban. 2009. *Cruising Utopia: The Then and There of Queer Futurity*. New York: New York University Press.

Munoz, Vic. 2012. "Gender/Sovereignty." In *Transfeminist Perspectives in and beyond Transgender and Gender Studies*, edited by Finn Enke, 23–33. Philadelphia, PA: Temple University Press.

Qwo-Li Driskill. 2016. *Asegi Stories: Cherokee Queer and Two-spirit Memory*. Tucson: University of Arizona Press.

Samuels, Ellen. 2017. "Six Ways of Looking at Crip Time." *Disability Studies Quarterly* 37(3). http://dx.doi.org/10.18061/dsq.v37i3.5824.

Schalk, Sami. 2018. *Bodyminds Reimagined: (Dis)ability, Race and Gender in Black Women's Speculative Fiction*. Durham, NC: Duke University Press.

Index

Page numbers in *italics* denote figures.

For Product Safety Concerns and Information please contact our EU representative GPSR@taylorandfrancis.com
Taylor & Francis Verlag GmbH, Kaufingerstraße 24, 80331 München, Germany

www.ingramcontent.com/pod-product-compliance
Ingram Content Group UK Ltd.
Pitfield, Milton Keynes, MK11 3LW, UK
UKHW020159200926
475702UK00004B/528
9780367086510